GENE STUDIES IN BIOLOGY

GENE STUDIES IN BIOLOGY

Edited by

Dr. Vijay Upadhe

Delve PUBLISHING

www.delvepublishing.com

Gene Studies in Biology

Dr. Vijay Upadhe

Delve Publishing
224 Shoreacres Road
Burlington, ON L7L 2H2
Canada
www.delvepublishing.com
Email: orders@arclereducation.com

ISBN: 978-1-77469-862-4 (Hardcover)

ABOUT THE EDITOR

Dr. Vijay Jagdish Upadhye is presently working as Deputy Director at Research and Development Cell (RDC) and as an Associate Professor, Department of Microbiology, Parul Institute of Applied Sciences (PIAS), Parul University, Vadodara, Gujarat. His area of research includes Immunodiagnostics for Tuberculosis, hepatitis, brucellosis and Infectious Diseases (more than 12 years). He has successfully contributed >76 publications on Tuberculosis, Infectious Diseases, Agriculture and other microbiological aspects. He has 17 chapters & 6 Books contributions, 40 IPR (26 Patents 9 Design Registrations and 5 copyrights) to his credit. He received grants from Gujarat State Biotechnology Mission (GSBTM), Govt of Gujarat for research on development of antigen detection test for early diagnosis of Brucellosis. Grant is also received from Students Open Innovation Challenge (SOIC) SSIP, Govt of Gujarat for research on development of antigen detection test for early diagnosis of Hepatitis B Infection. Till now, he has successfully completed two International and Three national projects. He has 13 several awards and notable accomplishments to his credit including Prof. G P Talwar Young Scientist Award, Senior Scientist Award 2020-2021 and Best Teacher Award 2019 (from Microbiologists Society India), Asian Pacific Federation of Clinical Biochemists (APCCB) Silver Jubilee Travel Fellowships, Best Leadership Award from Immunoshop India Pvt Ltd, Prof. R. H. Pandya Academic Excellence award, several best poster presentation award, and further more. He delivered >21 guest talks and was an organizing Secretary for Scientific events >11. He is working as technical consultant and knowledge partner with emerging diagnostic companies for development of diagnostic tools for Infectious diseases.

TABLE OF CONTENTS

LIST OF FIGURES

PREFACE

Human survival and that of other life forms with which we interact, directly or indirectly, depends critically on genetics. To obtain an understanding of the dynamics of life, it is crucial to study all areas of genetics and its applications. Since many biological processes are not controlled by gene mutations but rather by reversible and heritable epigenetic phenomena including DNA methylation, histone changes, and prions, epigenetics is often referred to as the "new or latest genetics." Many biological processes, including tissue/organ regeneration, X-chromosome inactivation, stem cell differentiation, genomic imprinting, and aging, are regulated by epigenetic mechanisms, which occur in a variety of animals. Numerous diseases, including cancer and diseases of the immunological, endocrine, and neurological systems, are caused by epigenetic abnormalities; some of these diseases have already received therapeutic treatment, and many more novel epigenetic therapies are probably just around the corner.

This book helps us comprehend the essentials of genetics and also discusses epigenetics and the basics of how genes function.

INTRODUCTION

The 100th Anniversary of the Genetics Society is a momentous occasion to celebrate the society's profound contributions to generations of geneticists throughout their research careers. Moreover, it is an opportunity to honor the discipline of genetics itself—a powerful and elegant field that has offered invaluable insights into the intricacies of life and its workings.

Genetics rests upon three fundamental pillars that underpin our comprehension of how living organisms function, evolve, and can be explored. The first pillar revolves around transmission genetics, which forms the foundation of heredity. Originating from Mendel's pioneering work and championed by William Bateson, the concept of the gene plays a central role in this aspect. Although Fisher's statistical analysis raised questions about the perfection of the experimental data, the significance of genes in inheritance remains undeniable.

The second pillar delves into how an organism's genotype dictates its phenotype. This involves understanding how genetic information stored in the gene influences the observable traits of an organism. Watson and Crick's seminal discovery in 1953 recognized DNA's role as a digital information storage device, determining protein sequences and consequently shaping the characteristics of phenotypes. The interplay of these two pillars offers valuable insights into the workings of living organisms and their evolutionary processes.

The focus of this article lies in the third pillar, which centers on harnessing genetics as a powerful tool to investigate biological processes. Genetic methodologies offer unique and profound ways to unravel the underlying molecular mechanisms of biological phenomena, even in cases where the mechanistic basis is initially unknown. Through these techniques, researchers can gain valuable insights into the intricate processes that govern life.

Diseases brought on by mutations and errors in the body's own molecules are currently the worst killers. A single gene abnormality that results in early death, or significantly altered bodily development affects about 5% of the global population. The likelihood that certain gene combinations may result in other diseases is high, yet it is very challenging to identify these. Since these issues typically affect species after the age of reproduction, when they have already passed along the genes that cause them, evolution has not done much to protect against them. New options for patients are made possible by new medical discoveries and technological advancements. Over the past few decades, as medical genetics became recognized as a medical specialty, new bioethical concerns have surfaced. The Human Genome Project and its funding was heavily invested in studying the moral, legal, and social ramifications of human genetics due to the importance and complexity of these concerns. Some of these ramifications such as genetic testing, gene therapy, and embryonic stem cell research have to be studied.

Few people in the medical field would argue against the creation of novel medications and treatments when a demonstrated advantage is realized. The knowledge of the human genome may be the one aspect of the medically relevant applications of genetic research that is challenging. There is a lot of discussion surrounding genetic screening and the potential for genetic discrimination. The primary ethical conundrums typically revolve around whether or not a fetus should be terminated if a disease-causing feature is found. The possibility of genetic discrimination based on knowledge of the genome sequence has created a new area of concern, where predisposition as well as actual circumstances that a person may experience become active concerns. When genotypes at two or more loci produce genotypic values that are more or less than the sum of the genotypic effects of the loci when evaluated separately, this phenomenon is known as epistasis. When there is interaction between two or more loci, physiological epistasis is present.

In conclusion, the Genetics Society's centennial celebration not only commemorates the society's contributions but also highlights the enduring significance of genetics as a discipline that continues to enrich our understanding of life's complexities and its inner workings. The power of genetics lies not only in understanding heredity and the relationship between genotype and phenotype but also in its capacity to unlock the secrets of life's underlying mechanisms.

Chapter

1

INTRODUCTION TO GENETICS

CONTENTS

1.1 INTRODUCTION

A young American geneticist named James Watson and his British colleague Francis Crick (Figure 1.1) had a brilliant insight into the structure of the DNA molecule in 1953 that accomplished several things (Portin, 2014). They demonstrated that DNA almost certainly contained the genetic material of cells and organisms. They revealed how cells could copy DNA to pass it along to their offspring. They showed how the molecule might change through mutations, which make evolution possible and demonstrated how DNA could copy itself to make mutations possible. Their paper also provided an explanation for research that quantified the four nucleotides, or DNA building blocks, in various animals. For fifty years, scientists had been aware of the existence of genes, but the substance that made them up remained a mystery.

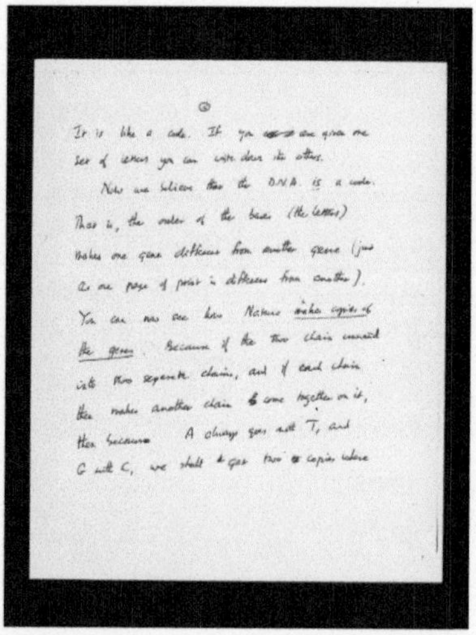

(3)

It is like a code. If you are given one set of letters you can write down the others.

Now we believe that the D.N.A. is a code. That is, the order of the bases (the letters) makes one gene different from another gene (just as one page of print is different from another). You can now see how Nature makes copies of the genes. Because if the two chains unwind into two separate chains, and if each chain then makes another chain come together on it, then because A always goes with T, and G with C, we shall get two copies where

Figure 1.1. Letter from Francis Crick to his son, Michael, explaining his and Watson's discovery of the structure of DNA. The letter is the earliest written description of the genetic mechanism on 19 March 1953. Credit: Wellcome Library, London. (Source: Tamura K. The Genetic Code: Francis Crick's Legacy and Beyond. *Life*. 2016; 6(3):36. https://doi.org/10.3390/life6030036, Creative Commons License).

That question was instantly answered by Watson and Crick's research, but many more were also brought up. DNA was handed from parent to child, or from plant to seed, and it also actively participated in a cell's everyday activities. The American geneticists George Beadle (1903-89) and Edward Tatum (1909-75) proved this in the 1940s (Raju, 1999). Their research demonstrated that a single gene mutation resulted in a form of mold losing a particular enzyme. This clearly implied that each gene, whatever its constituent parts, produced a single protein. Proteins and DNA are entirely separate classes of substances.

Figure 1.2. Rosalind Franklin. (Source: By The Editors of Encyclopaedia Britannica. "Rosalind Franklin, British Scientist," Encyclopaedia Britannica (article created 20 Jul 1998, accessed 29 Mar 2021); https://www.britannica.com/biography/Rosalind-Franklin : https://cdn.britannica.com/30/99730-050-E68F62FF/Rosalind-Franklin.jpg, Fair use, https://en.wikipedia.org/w/index.php?curid=24959067, Creative Commons License).

When Crick announced what is referred to as the central dogma of molecular biology in 1958, "DNA produces RNA makes proteins", he gave us a path toward discovering the solution. In other words, the genetic code was translated into RNA, an intermediary molecule that was employed to create proteins. Nobody knew how that occurred; Crick's notion posed a puzzle to the scientific community as a whole. Researchers from all over the world figured out the basic steps in about 15 years. Naturally, the answers generated new queries that, in one way or another, constitute the focus of the majority of the research being conducted in biological laboratories today. Although each RNA was created from the data contained in a gene, it was unable to convey information back and alter the gene. Similar to how an RNA molecule can be used to create a protein, an RNA cannot be altered by a protein. For instance, RNAs can occasionally rewrite the information contained in genes. One gene produces one enzyme, according to Beadle and Tatum's theory, and has also been shown to be overly simplistic. Although a gene cannot make two entirely distinct proteins, one RNA can be used to produce two quite different proteins, much like how the same components and recipe can produce various dishes if a cook omits, adds, or reorders certain processes. Cells have developed control mechanisms that allow them to intervene and reroute or obstruct the process at any stage in the conversion of genetic information into proteins.

Most scientists consider how genes, RNAs, proteins, and other components interact to determine what occurs within cells. Going deeper and looking at these molecular structures is crucial to comprehend how they act. It is similar to an engineer's job. Several cell components are necessary for replicating it or fixing it. For one component to be able to be coupled with others, it must have the proper size and shape. Molecules contain various layers of structure, all of which are crucial to their operations, similar to how a motor has huge parts that are composed of smaller parts. Atoms that interact with one another to create bonds make up biological compounds. Protons and neutrons make up the core nucleus of an atom, which is surrounded by electrons (Langanke et al., 2003). Early atomic models portrayed this as a miniature solar system, but in the early 20th century, a new branch of physics known as quantum mechanics caused this to shift. In the new atomic model, electrons are placed in zonal orbits around the atom that are referred to as shells. These zones are spread out over a wide area around the nucleus. Only a certain number of electrons can fit inside each shell. Two electrons can fit in the innermost shell, and eight can fit in each of the following two. However, the shells are frequently incomplete, lacking one or more electrons. Atoms want to fill their empty shells by joining up with other atoms and exchanging electrons with them since they cannot be stable until their shells are entirely filled.

1.2 BUILDING BLOCKS

Hydrogen is the most basic atom. It only has one electron; hence it lacks one in its outer shell, which houses two electrons. Bonds between hydrogen and atoms without an electron in their own outer shells are simple to form. Since helium has two electrons, it can exist alone and does not require fusion. This makes it similar to neon, argon, and a few other noble gases, all of which have full outer shells and are considered inert because they are slow to react in chemical or biological processes. When compared to, say, the behavior of oxygen, this is very different. Eight electrons make up oxygen, which are divided into two shells: two in the inner shell and six in the outer one. By joining forces with two hydrogen atoms to form the molecule written as H_2O, or water, it may quickly fill those gaps. Covalent bonds are formed between atoms when they adhere to one another by exchanging electrons (Peterson & Treagust, 1989). Covalent bonds can also connect groups of atoms (molecules) together, but as a group gets bigger, its overall shape becomes more crucial to what it can bind to. To share electrons, atoms must be close to one another. Sometimes the bonds in a molecule prevent atoms from interacting with one another, rendering the molecule inflexible. Every biological process depends on the creation and dissolution of chemical bonds. Ionic bonds, which occur when one atom takes an electron from another rather than sharing it, and van der Waals interactions, which are non-chemical forces between atoms and molecules, are examples of alternative non-covalent methods for one atom to bind to another. However, in general, covalent bonds play a major role in the organization of biological molecules, including the structures of proteins, DNA, and RNA. Due to the body's high water content and the ease with which this element can be incorporated into molecules, they frequently involve hydrogen atoms.

Each DNA, RNA, and protein exists in its most basic form as a chain of smaller molecules. Amino acids are considered as the building blocks of proteins. 20 different types of amino acids have been created by evolution, which are strung together in cells by other molecules or obtained from food (Li et al., 2019)regio-, and chemoselective enzymes has enriched the toolbox of synthetic organic chemistry. Among the different gene mutagenesis techniques, saturation mutagenesis (SM. All 20 types of amino acids are required by every organism, however, not all organisms can produce all of them. Leucine, for instance, cannot be produced by human or animal cells; therefore, people must consume it through food. The bacterium that causes tuberculosis, on the other hand, is unable to get leucine externally and must produce its own. Researchers are currently working on medications that will prevent the bacteria from producing leucine. The bacterium might be killed by such a drug without endangering the cells it has infected. Some people claim that a laboratory strain of tuberculosis that is unable to make leucine does not proliferate when it infiltrates human cells. Blocking just one process in the leucine synthesis machinery might be sufficient to lessen the bacterial strength. Amino acids are atoms arranged in tiny clusters. Each type contains a carbon atom at its center, which is bound to other atoms on four sides. There is always one hydrogen atom in one direction. A structure known as an amino group can be found on the second side. A -COOH or carboxyl group, consisting of one carbon atom (C), two oxygen atoms, and one hydrogen atom, is present on the third. Each amino acid is different due to the side chain, a collection of atoms that originates on the fourth side.

1.3 AMINO ACIDS

Similar to how the letters in this text are connected to one another, a protein is made up of a lengthy chain of amino acids (Figure 1.3).

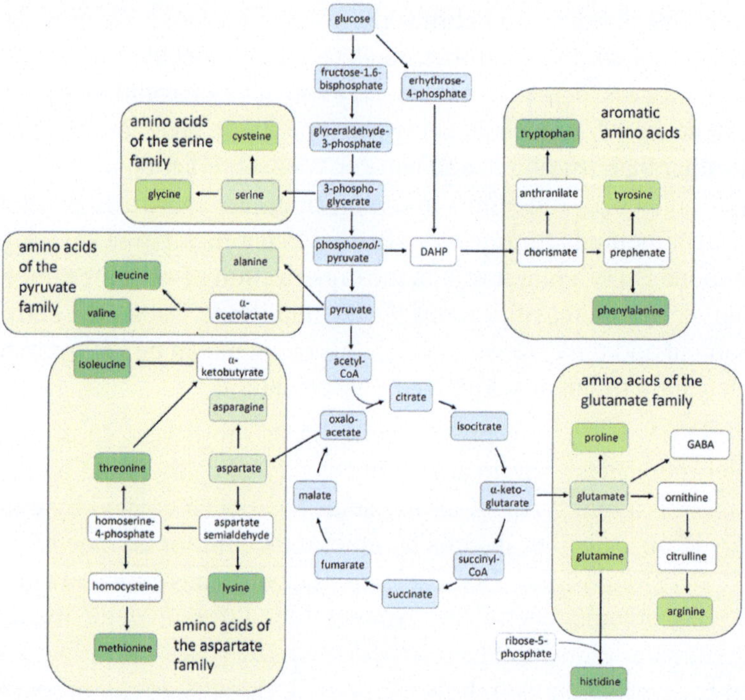

Figure 1.3. According to their respective precursors, the amino acids are grouped into five families derived from glutamate, serine, pyruvate, aspartate, or chorismate. The nine amino acids that cannot be synthesized in animals are shown in dark-green boxes, while those that can be synthesized but additionally need to be taken up with the diet are in brighter boxes. Proteinogenic amino acids that can be sufficiently synthesized in animals are in pale green boxes and non-proteinogenic amino acids and other important intermediates are boxed in white. DAHP, 3-deoxy-D-*arabino*heptulosonate-7-phosphate. (Source: Trovato M, Funck D, Forlani G, Okumoto S and Amir R (2021) Editorial: Amino Acids in Plants: Regulation and Functions in Development and Stress Defense. Front. Plant Sci. 12:772810. doi: 10.3389/fpls.2021.772810, Creative Commons License).

When the carboxyl group of one amino acid joins the amino group of the next, two amino acids are said to be bound. A water molecule is released during the binding process, which also results in the establishment of a peptide bond. The amino acids can release one another again by adding water. Enzymes that dissolve the bonds and liberate the molecular constituent elements to engage in novel interactions can quicken the process.

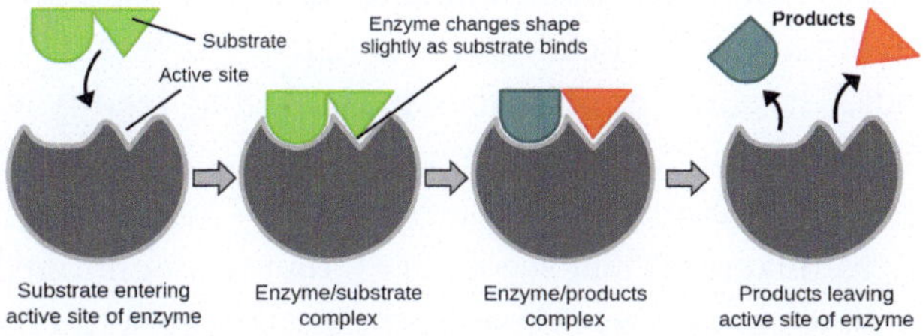

Figure 1.4. Enzymes. (Source: By Created by TimVickers, vectorized by Fvasconcellos - Provided by TimVickers, Public Domain, https://commons.wikimedia.org/w/index.php?curid=1267900, Creative Commons License).

Enzymes (Figure 1.4) control the chemistry of the cell in this way. By instructing it on which amino acids to use and how to combine them, the information in a gene teaches the cell how to produce a protein.

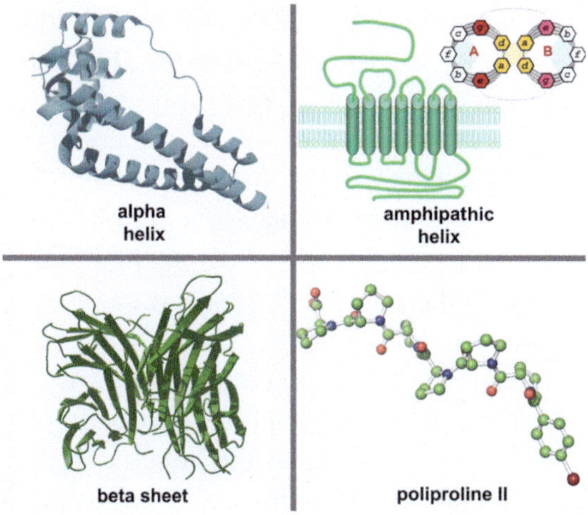

Figure 1.5. Alpha helix and beta strand. (Source: Caporale A, Adorinni S, Lamba D, Saviano M. Peptide–Protein Interactions: From Drug Design to Supramolecular Biomaterials. *Molecules*. 2021; 26(5):1219. https://doi.org/10.3390/molecules26051219, Creative Commons License).

After that, amino acids join together to form chemical connections that can occasionally be assisted by partner molecules to produce a variety of folds and structures. A protein's primary structure is a string of its amino acids joined together by peptide bonds. The secondary structure develops as a result of chemical attraction between nearby amino acids in the string, which results in the formation of tiny folds (Fülöp et al., 2006).

This typically results in either an alpha helix or a beta strand, two coil forms (Figure 1.5). Larger beta sheets frequently contain links between adjacent strands. The surfaces of helices and coils may readily attach to other regions of the protein since their atoms similarly lack electrons. Secondary structures are typically connected to one another by loose pieces of the string known simply as linkers. When secondary structures pull together and fold into bigger modules known as domains, the tertiary structure is created. As a result, some amino acids are kept inside while others are left exposed on the exterior where they can interact with other molecules or domains. Such modules or moieties may be found within a protein, each joined by a few weak connecting strands. This gives the protein a unique three-dimensional form that is crucial to how it works. These structures are frequently rigid and can only be altered by physical force or a significant alteration in the chemistry of the relevant domain, which breaks certain bonds and permits others to form.

A full protein's outer surface, which includes all of its domains, can attach to other molecules to form its quaternary structure. The majority of the work in a cell is done by machines made of numerous proteins or combinations of proteins, RNA, DNA, or other molecules. The core amino acid string's form and chemistry change as it folds into these higher levels of structure, becoming so complicated that it can occasionally bind to just one partner. The primary reason that tens of thousands of distinct molecules are able to organize themselves into the precise, durable structures that make up a cell is because folding prevents the wrong molecules from mixing. There are 11 key proteins found in an exosome: six of the proteins, which dock onto one another to create a ring as its core, are present (D. Wang et al., 2021). This ring's chemistry enables it to dissociate the bonds tying together the RNA nucleotides. The other exosome proteins assist in positioning and guiding an RNA into the core where it comes into contact with the ring proteins that break it down. Some proteins form extremely stable machines when they bind to one another for extended periods of time. Others just connect temporarily to complete a task; after that, the machine is disassembled once more and its parts are put to work again. It was only in 2002 that researchers in Heidelberg, Germany, realized just how dynamic this condition is. For instance, the protein beta-catenin has the function of passing instructions to other genes. When they become active, the cell makes new molecules that have the power to significantly alter its behavior. Beta-catenin typically needs to be kept off since the instructions can only be sent at specific times.

The cell controls this by joining it with other molecules. These compounds both summon up additional machinery that breaks down beta-catenin and stops it from getting to the nucleus, which is what it needs to do to activate genes. Beta-catenin is delivered to the nucleus by other molecules when the behavior of the cell needs to alter (Mikami et al., 2000). Gene mutations can be harmful because they frequently result in modifications to various levels of protein structure. Changing one DNA letter frequently results in changing one amino acid letter in a protein's main structure. This could modify the close-range interactions between amino acids, changing how alpha helices and beta sheets are formed (secondary structure), how domain shapes are disrupted (tertiary structure), and which molecules the protein can bind to. The result is similar to using a part meant for one model of car in another. If it doesn't fit, the component could shatter, potentially wrecking the engine and even the entire car.

Figure 1.6. Alois Alzheimer. (Source: https://upload.wikimedia.org/wikipedia/commons/a/ad/Alois_Alzheimer_003.jpg, Creative Commons License).

Peter T. Landsbury, a structural biologist at Harvard Medical School, Christopher Dobson, of the University of Leeds, and other labs conducted a number of studies in the 1990s that demonstrated that various mutations can have a similar impact on protein structure by converting alpha helices into beta strands, sheets, and finally tight clumps that are incapable of being broken down by the body. Alois Alzheimer (1864–1915), a German physician, (Figure 1.6) made the first discovery of these clumps in the brain of a woman who had passed away from a degenerative disease that had rendered her mentally incapacitated (Dahm, 2006). He discovered that the brain's nerve cells had died and developed signs of Alzheimer's disease because dense amyloid fibers, which are primarily formed of beta sheets, had accumulated between the cells.

1.4 IMPORTANCE OF PROTEIN STRUCTURES

Since then, over 20 distinct neurodegenerative diseases have been linked to fiber-forming proteins. The study of the architecture of proteins and other molecules encompasses a whole field called structural biology. The discovery and refinement of new drug classes is a crucial application of this research. The majority of drug active components are often tiny compounds that alter the shapes and functions of proteins by docking onto them (Figure 1.7).

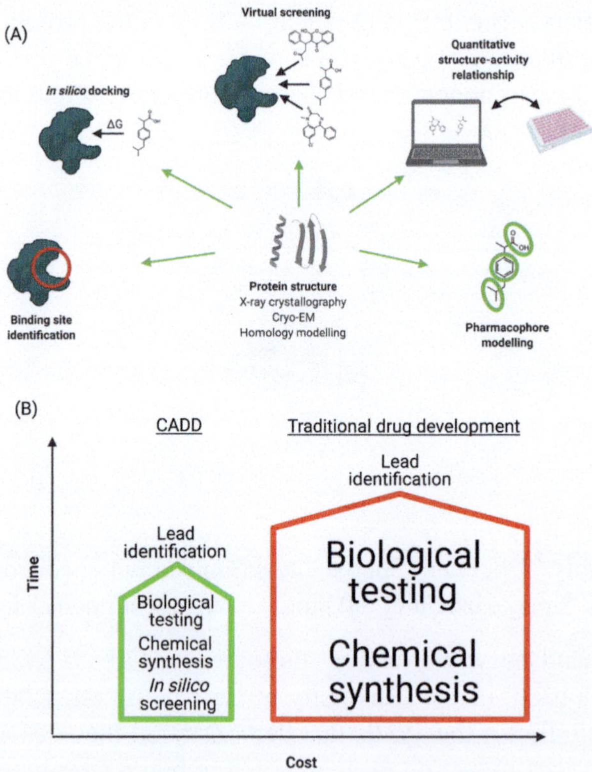

Figure 1.7. Drug docking. (Source: Salman MM, Al-Obaidi Z, Kitchen P, Loreto A, Bill RM, Wade-Martins R. Advances in Applying Computer-Aided Drug Design for Neurodegenerative Diseases. *International Journal of Molecular Sciences*. 2021; 22(9):4688. https://doi.org/10.3390/ijms22094688, Creative Commons License).

For instance, a protein surface can possess a pore that must be filled by a tiny molecule for the protein to be activated. A drug that seals the pole prevents this from happening. Aspirin functions in this way. Since 1899, when the Bayer corporation started marketing the drug internationally, it has been in use. But until 1971, its functions remained a mystery.

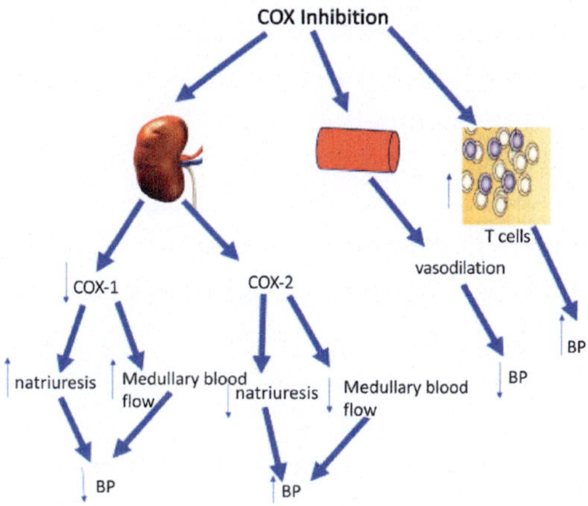

Figure 1.8. COX-1 and COX-2. (Source: Khan S, Andrews KL, Chin-Dusting JPF. Cyclo-Oxygenase (COX) Inhibitors and Cardiovascular Risk: Are Non-Steroidal Anti-Inflammatory Drugs Really Anti-Inflammatory? *International Journal of Molecular Sciences*. 2019; 20(17):4262. https://doi.org/10.3390/ijms20174262, Creative Commons License).

In that year, John Robert Vane (1927-2004), a researcher at the Royal College of Surgeons in London, found that asprin inhibits sites in the proteins COX-1 and COX-2 (Figure 1.8) that are necessary for the creation of fat molecules called prostaglandins (Patrono, 2023). Structural biologists may identify the precise areas where molecules hook onto one another by analyzing the map of a protein. They can also ascertain what properties a drug would need to possess in order to disrupt a protein's function. They must first establish a three-dimensional structure of the molecule, which is challenging to perform. Structural biologists would wish to learn how to anticipate what secondary, tertiary, and quaternary structures a protein would form based on its amino acid notation because it is much simpler to determine a molecule's sequence. If this were the case, it would be able to design and create artificial molecules with specific shapes intended to regulate the behavior of other molecules. A mutation's impact on the structure and function of a protein might potentially be observable.

Small molecules can now be designed with specific forms, but researchers have not yet figured out how larger molecules acquire their folds. Some proteins have thousands of amino acids, and each one has the ability to bind to numerous other amino acids. Even the most powerful computers of today are unable to predict which folds will be created by the chemical forces of all the amino acids since there are so many potential folds in a single molecule. However, there may be ways to find shortcuts; this is a topic of intense investigation. In reality, the American National Institutes of Health hold an online competition with this theme every two years. A string of amino acids is provided to laboratories throughout the world, and they use their computer programs to anticipate what shape it will take. Within two days, the calculations are completed entirely by the computer. Although DNA and RNA are constructed of distinct nucleotide subunits, the same fundamental laws of chemistry govern how they work. No matter if they exist as solitary organisms in a pond or as components of a massive human body, cells must be adaptable to survive. They must travel toward food sources, react to environmental changes, and defend themselves from the environment's harmful elements or predators. These cell actions are the result of communication between the environment and the data stored in its DNA. Over the course of evolution, the genome's instructions have enabled the species to persist. The cell tries a new set of instructions as conditions change. It alters its proteins and reassembles molecular machinery by turning off some of the genes it had been using and turning on a different set. The environment typically causes changes in the cell.

1.5 ENVIRONMENTAL SIGNALS

Environmental cues are integrated to influence how cells behave (Newman, 1999). Each type of receptor protein found in cells can recognize a distinct signal through a ligand receptor molecule. A distinct collection of receptors is produced by each type of cell, allowing it to detect some substances but not others. The linings of blood vessels, the bone marrow, and the brain all create different ligands than do the muscles of the heart. Cells learn how to grow, when to divide, and even when to die by deciphering combinations of signals. The first stage of a process that instructs a cell on what to become and how to behave is sensing. Receptors use signaling pathways to transmit information to cells. These are information chains consisting of molecules that occasionally come into contact with one another and bond briefly enough to alter one another's chemistry. Through signaling pathways, specific molecules are activated in a predetermined order, or cascade, which results in the activation of specific genes. DNA contains the instructions for additional molecules that the cell needs (RNAs and proteins). Signaling pathway information turns on some genes while turning off others. Signaling pathways have a role in almost all cellular events, both positive and harmful (Garrington & Johnson, 1999)in part, by use of scaffolding and anchoring proteins. Scaffold proteins bring together specific kinases for selective activation, sequestration and localization of signaling complexes. The recent elucidation of scaffolding mechanisms for MAPK pathways has begun to solve the puzzle of how specificity in signaling can be achieved for each MAPK pathway in different cell types and in response to different stimuli. As new MAPK members are defined, determining their organization in kinase modules will be critical in understanding their select role in cellular regulation.","container-title":"Current Opinion in Cell Biology","DOI":"10.1016/S0955-0674(99. A protein may vanish or become stuck in transmitting mode, broadcasting signals continuously even when the receptor has not communicated. Cancer may develop if the route instructs the cell to divide continuously rather than alternating between the reproduction and resting phases. Whether cells adhere to one another in a tissue or migrate is one mechanism regulated by signaling pathways. The ability to turn on and off migratory activity was crucial for the animal evolution more than a billion years ago, and it is still crucial for every animal alive today. Serious issues, like cancer and metastases, may emerge when the signals that regulate this action fail.

1.6 CANCER AND SIGNALS

Cancer is similar to what occurs frequently in an embryo: In order to create new tissues, cells move to certain locations. Tumors resemble the body's unproductive attempts to develop additional organs that have no purpose and are not the result of evolution. Other molecular processes that cells require are triggered by signals. The architecture of the cell is being constructed, which is important for its behavior. For instance, a neuron's tree-like structure, which has extensive networks of roots and branches, enables it to interact and communicate with other cells. For a red blood cell to pass through capillaries, which are minuscule blood vessels, it must be small and flexible. Piston-like fibers are necessary for muscles to be able to expand and contract. Cells have unique forms because of protein fibers and tubes. A cell also needs to transform environmental materials into what it needs in order to survive and divide. Protein and RNA-based machinery breaks down and reassembles raw materials to produce energy and nutrition. Many of the instruments used to research proteins and other molecules were originally created for physics as tools to study atoms or the atomic scale behavior of energy. The use of X-rays to study the structure of biological molecules has been one of the most significant advancements (Campbell, 2002). Doctors and others employed X-rays for several decades after the German physicist Wilhelm Roentgen (1845-1923) made the discovery without fully understanding what they were (Flynt, 2020). Some scientists believed that X-rays were waves, but German physicist Max Laue (1879–1960) claimed they were streams of particles (Busch, 2021). To solve the problem, Laue devised an experiment employing crystals.

The majority of the energy in X-rays that are directed through a crystal simply passes through. However, some of the rays in the sample are diffracted, or are radiated at an angle, as a result of collisions with the electrons in the atoms. On a photographic plate or a detector, these waves are recorded. A molecule must be arranged in a fairly regular structure, like the identical unit cells of a crystal, in order to produce a meaningful pattern. The first X-ray images of molecular structures were taken of salt and minerals, which are composed of a few atoms grouped in basic shapes. The 1915 Nobel Prize in physics was shared by the British physicists Sir Lawrence Bragg (1890–1971) and Sir William Bragg (1862–1942) for this achievement (Jenkin, 2001).

Figure 1.9. Linus Pauling. (Source: Glendening ED, Weinhold F. Pauling's Conceptions of Hybridization and Resonance in Modern Quantum Chemistry. *Molecules*. 2021; 26(14):4110. https://doi.org/10.3390/molecules26144110, Creative Commons License).

Their work paved the way for young scientists who had worked in the Braggs' labs to explore the structural properties of more complex compounds in newly established crystallography departments at British universities. Alpha helices and beta sheets are the fundamental secondary structures of proteins, which Linus Pauling (1901–94), an American chemist (Figure 1.9) at the California Institute of Technology, discovered using his data. It was a significant achievement that contributed to Pauling winning the Chemistry Nobel Prize in 1954. John Bernal (1901–1971), a worker in William Bragg's London laboratory, relocated to Cambridge University and established a crystallography department there. When Dorothy Crowfoot Hodgkin (1910–1994), one of his students, received a sample of the hormone molecule insulin in crystal form, she developed an interest in the potential for employing X-rays to analyze proteins. The first X-ray photographs of the crystal were taken by Bernal and Hodgkin in 1934, but they were unable to translate their findings into a model of the structure of insulin. However, the protein crystals generated distinct diffraction patterns, giving the researchers optimism that they might one day be used to figure out how the atoms in molecules are arranged.

Figure 1.10. Max Perutz. (Source: https://upload.wikimedia.org/wikipedia/commons/8/8b/Max_Perutz_1962.jpg, Creative Commons License).

Max Perutz (1914-2002), (Figure 1.10) another of Bernal's pupils, had grown interested in protein structure. Perutz arrived in Cambridge from Austria in 1936 after earning a degree in inorganic chemistry there. When the Nazis occupied Austria and Czechoslovakia, he fled the nation, and his family fled as well. Due to suspicions raised by Perutz's nationality when war broke out between Britain and Germany, he was briefly detained. Perutz had learned about the contributions made by Linus Pauling and other field pioneers. He obtained crystals of the hemoglobin protein. He wanted to utilize these crystals to determine the structure. Perutz started a 22-year endeavor to obtain a structural map of hemoglobin after Bernal taught him how to use the X-ray equipment. A British student named John Kendrew (1917–1977) joined him in 1947 and rapidly developed a similar level of enthusiasm for protein structure. Proteins include 20 amino acids, as opposed to a single atom like a diamond or the repetitive structures of an artificial fiber, making them chemically complex. The patterns created by X-ray studies were so intricate that it seemed impossible to understand them. It is simplest to obtain diffraction patterns from heavy atoms with lots of electrons, like those found in metals, because X-rays interact with their electrons. The majority of the atoms in biological compounds are light atoms with few electrons. Diffraction patterns also failed to record phase; a sort of data essential for matching pattern spots to atoms in molecules. Phase has to do with the possibility of interference between two X-ray waves that were emitted by a protein crystal. Typically, they neutralize one another. The diffraction pattern is created by the escaping X-rays.

Photographic dots on the plates captured a wave's intensity but not its phase. Perutz and Kendrew found a way to fix these issues by soaking protein crystals in heavier elements, like mercury, that would bond to the side chains of particular proteins (Jaskolski et al., 2014). Strong, distinctive signals emitted by these atoms could be used to pinpoint and map the locations of other atoms. By contrasting the patterns created by proteins that have been soaked in atoms with an image from crystals that have not been soaked, they could also be utilized to determine the phase. Perutz stated in his presentation that discovering the structure was a first step toward a bigger objective: elucidating how the protein's structural design enabled it to transport oxygen.

Nowadays, protein crystallography is the primary technique utilized to reveal the structures of proteins. Synchrotrons, which are ring-shaped tunnels used to collide atomic particles, have taken the role of the X-ray machines used in laboratories in the 1950s and 1960s (Potts et al., 2005). The British scientist Ken Holmes, who had studied under Franklin, conducted a number of studies at a synchrotron in Hamburg, Germany, in 1969, demonstrating that protein structures could be determined using the high-energy X-rays produced by these devices. Another technique that was adopted from physics and is now widely used to show the structures of proteins and other molecules is nuclear magnetic resonance (NMR) (Figure 1.11). High quantities of proteins must be drawn out of a liquid and gently dried until they bind to one another in symmetrical configurations in order to form a crystal. It is impossible to investigate every protein structure using this method or to address every research concern involving proteins. Not all proteins can grow crystals of sufficient quality for research. Highly synthetic forms of molecules are frequently seen in crystals.

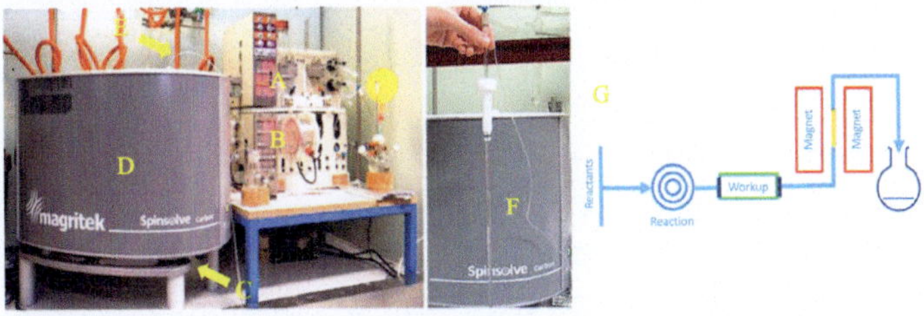

Figure 1.11. An example of an NMR system used in a lab. (Source: Yu H-Y, Myoung S, Ahn S. Recent Applications of Benchtop Nuclear Magnetic Resonance Spectroscopy. *Magnetochemistry*. 2021; 7(9):121. https://doi.org/10.3390/magnetochemistry7090121, Creative Commons License).

1.7 NMR

Isidor Rabi (1898-1998), (Figure 1.12) a physicist at Columbia University, discovered the fundamentals of NMR in the 1930s.

Figure 1.12. Isidor Rabi. (Source: By Nobel foundation - http://nobelprize.org/nobel_prizes/physics/laureates/1944/rabi-bio.html, Public Domain, https://commons.wikimedia.org/w/index.php?curid=6183671, Creative Commons License).

He was researching the forces holding electrons to atomic nuclei using molecular beams; this study won him the 1944 Nobel Prize in Physics. It had useful applications in the creation of radar, a project Rabi was hired to work on during World War II. Felix Bloch (1905-83) and Edward Purcell (1912-97), two physicists, modified NMR so that it could be used to examine liquids and solid things. Their efforts helped them win the 1952 Physics Nobel Prize. Kurt Wuthrich made significant contributions to the development of NMR as a technique for determining the structures of biological molecules over time (Wüthrich, 2001). When Wuthrich joined Bell Laboratories in Murray Hill, New Jersey, in 1967, he was given access to one of the best instruments available at the time. This is where he first started working with NMR. He utilized it to research the composition and behavior of proteins in liquids; partly as a result of his work, it is now a common technique in the fields of structural biology, drug development, and related disciplines. His achievements were honored in 2002 when he received the Nobel Prize. Finding individual atom positions and determining their positions in relation to other atoms within the same molecule are necessary to obtain a molecular structure. The nuclei of some atoms absorb energy that can be precisely measured when exposed to a magnetic field of a specific strength. Energy measurements can identify specific atoms since different types of atoms need different intensities of fields to operate in this way. The magnetic field must be relaxed once more in order to determine their positions in relation to one another. This result is comparable to placing a powerful magnet near a compass and then taking it away. The magnetic field drags the compass needle when it is nearby. The needle moves back to its initial place when the magnet is once more removed. In an NMR experiment, the atom nuclei experience the same outcome. NMR can only reveal details about discrete molecular areas. Due to this restriction, it typically cannot offer the structure of a whole molecule. However, it can study proteins in liquids, which is a state that is quite similar to their native forms. This means that during procedures like the binding of a chemical to a protein, researchers can also see changes in its structure.

1.8 MRI

The medical approach known as magnetic resonance imaging (MRI) (Figure 1.13) is based on the same ideas and methods (Mori & Barker, 1999)which has intriguing properties depending on the physiological and anatomical environment of the organisms studied. We explain the principles of this emerging technique and subsequently introduce some of its present applications to neuroimaging, namely detection of ischemic stroke and reconstruction of axonal bundles and myelin fibers. Anat Rec (New Anat.

In contrast to how an X-ray machine works, which detects mineralized materials like bone, an MRI detects the presence of specific fluids. This makes it possible to investigate the composition and functioning of the brain and other organs using it. The development of new techniques to look into the physical properties of proteins and other molecules occurred in the 1980s and 1990s. One of these, known as atomic force microscopy (AFM), uses physical contact to examine the shapes and properties of molecules, cells, and surfaces. Working at IBM in Zurich in the early 1980s, Gerd Binning and Heinrich Rohrer created a device that employs a microscopic needle to sense surfaces. Similar to an ancient vinyl record player with a very fine needle attached to an arm, the technology operates in a similar manner. This needle's tip moves across the grooved surface of a record as it moves along it. This motion is translated into sound by the machine. The needle is much smaller with an atomic force microscope; its tip is only one atom in size. It isn't literally pushed down to the surface when it is transported; rather, it is carried near enough to interact with the atoms there. When the two objects are very near together, the electrons of surface atoms jump, moving the needle's tip.

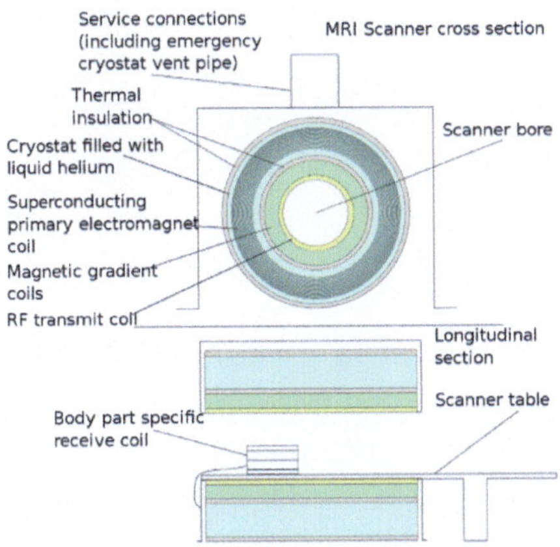

Figure 1.13. MRI. (Source: CC BY-SA 3.0, https://commons.wikimedia.org/w/index.php?curid=128258002, Creative Commons License).

The motion can be recorded by using sensors or a laser to follow the arm's movement. The data is put together by a computer into an atom-by-atom representation of the surface. The device has a resolution that is substantially higher than that of light or electron microscopes. It has been used to probe DNA and protein complexes as well as to create precise maps of the cells. Early in 2008, the technique was employed by Daniel Müller's biophysics group at the University of Technology in Dresden, Germany, to examine prions-infected cells. Mad cow disease and other neurological disorders are brought on by misfolded prion proteins. Prions form tough fibers in the test tube that do not degrade, but this process has not been directly seen in cell cultures or animals. AFM was utilized by Müller's team to examine the infected mouse brain cells (Schubert et al., 2018). They discovered that long, prion-containing fibers started to develop on the cells, confirming that cells behaved the same way as prions did in the test tube.

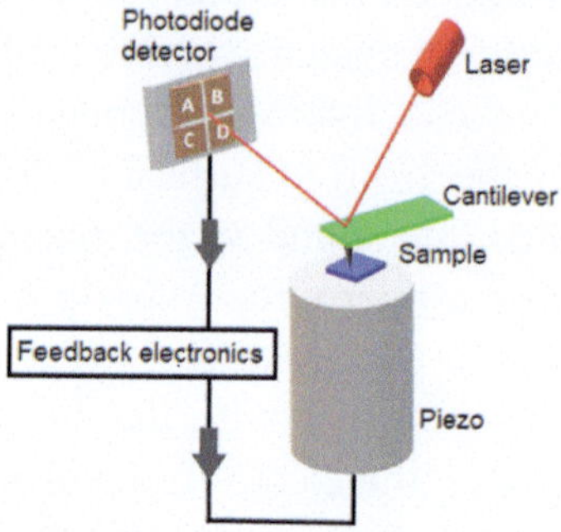

Figure 1.14. Scheme of AFM. (Source: Mišić Radić T, Vukosav P, Čačković A, Dulebo A. Insights into the Morphology and Surface Properties of Microalgae at the Nanoscale by Atomic Force Microscopy (AFM): A Review. *Water*. 2023; 15(11):1983. https://doi.org/10.3390/w15111983, Creative Commons License).

1.9 AFM

AFM (Figure 1.14) has also been used to examine the physical forces at play within protein structures and how they are affected by external forces (Haase & Pelling, 2015). Some molecules need to be stretched or go through other significant shape changes in order to operate. For instance, the protein titin aids in the contraction and relaxation of muscles. Sarcomeres, the structurally piston-like subunits that make up muscle fibers, are what give them their name. To slide the piston in and out, a sheath of fibers slides along a thicker fiber. The sarcomere's parts are held together by a long, string-like molecule called titin, which also permits them to expand and contract. When titin is forced to stretch, physical pressure causes some chemical links in the modules to break, causing them to unfold. The bonds are restored and the modules refold when the molecule relaxes once more. It was also believed that another part of titin called PEVK, which is more string-like also acts as a spring, but since the process had not been directly observed, it was hard to determine which sections were exerting the most force. Hermann Gaub and his team at the Institute for Applied Physics in Munich, Germany, made the decision to employ AFM to look into this issue in 1997. They connected one end of a titin molecule to a gold plate and then used the atomic force microscope's tip to close the other end. The team measured the force required to move the tip and the length of the molecule while it was being dragged. The outcome was a sawtooth-like pattern in which the force suddenly relaxed and the molecule rapidly became longer. As the dragging proceeded, the force increased once again; the titin molecule grew slightly longer; and then there was a subsequent decrease in force. This repeatedly occurred. Although it was difficult to be certain, Gaub believed that this indicated the unfolding of a single Ig domain.

Each domain behaved slightly differently and required a different amount of force to unfold because of even minute changes. The results of the tests demonstrated that AFM might be used to gather new kinds of data regarding the mechanical forces and structures that control the behavior of molecules. Similar to AFM, the main idea is to use a tiny probe to look at surfaces, molecules, and forces at the atomic level. In this instance, the probe is a tiny plastic ball rather than a needle, and it is held in position by a laser beam as opposed to being put on an arm. The development of photonic force microscopy (PFM) began with a series of tests conducted at Bell Laboratories in New Jersey in the 1970s (Kalinin et al., 2007). A novel method for manipulating objects at the nanoscale using lasers was being developed by researcher Arthur Ashkin. After many years of effort, a team at Bell Laboratories led by Stephen Chu used the method to successfully control and cool atoms. By the middle of the 1990s, Stelzer and his associates were employing the technique to develop a new kind of microscope. They were able to see the bead move by covering it in antibodies and attaching it to other proteins. The precise motion of the bead might be captured and used to recreate the motions of the proteins and the forces acting on them. The majority of the lipids that make up these envelopes enclosing cells form a double layer, resembling two soap bubbles sandwiched together with very little space in between. Such proteins as receptors, which take signals from the environment, float in the layers. Prior to the 1990s, the majority of scientists thought that proteins and lipids floated somewhat freely and independently of one another. A new model was then put up by Elina Ikonen of the National Public Health Institute of Finland and Kai Simons of the Max Planck Institute of Molecular Cell Biology and Genetics (Dresden). They made the observation that proteins are bound to lipids in the inside of the cell. Some lipids and proteins continue to be connected to one another when they are delivered to the membrane. Simons and Ikonen proposed that lipids such as cholesterol served as platforms, similar to tiny floating rafts, on which various proteins might be put together.

An excellent example is the skin, which has two sides: one that is in contact with the environment and the other that does so with the interior of the body. These two surfaces require various proteins and lipids because their functions are dissimilar. Horber and Stelzer came to the conclusion that the theory might be tested using photonic force microscope (PFM). Their teams connected the bead to a protein in the membrane that was assumed to be contained in a raft and used a photonic force microscope to observe its movement. They observed that because raft proteins were bound to lipids, they did not move as easily in the membrane. The proteins were liberated and permitted to move much more freely in another experiment that reduced the amount of cholesterol in cells, the glue that ties rafts together. Motor proteins, which act as the cell's drivers, have also been the subject of PFM research. Microtubules are the fibers that crisscross the cell. Microtubules and motors are essential for cell division as well as other functions, and they aid in constructing and altering the cell's architecture. Kinesin and other motor proteins move over the microtubule surface. With the creation of techniques that have offered scientists new ways to observe the behavior and activity of molecules, the light microscope is currently undergoing development. This is taking place despite the fact that light microscope resolution has not changed much since Joseph Lister (1786-1869) made some significant improvements to their design and lenses in the early 19th century. A light microscope can magnify items by around 2,000 times under ideal conditions (Humphreys, 2014). The electron microscope's resolution allows it to magnify objects 2 million times, yet even then it would not be able to distinguish individual proteins or other molecules. However, the discovery of fluorescent proteins and the methods for using them to investigate biological processes have led to a rebirth of light microscopy in the molecular age.

1.10 PROTEINS: INTERESTING RESEARCH

A species of jellyfish called Aequorea victoria (Figure 1.15) was used by Osamu Shimomura and his colleagues at Princeton University in New Jersey to isolate two flourescent proteins (Shimomura, 2005). The researchers had mastered their procedure by the middle of the 1970s and were gathering more than 3,000 jellyfish per day.

Figure 1.15. *Aequorea victoria*. (Source: By Sierra Blakely, Attribution, https://commons.wikimedia.org/w/index.php?curid=4091901, Creative Commons License).

These and other molecules, such as the luciferase proteins, were of great interest because they could convert chemical energy into light. Because jellyfish could only produce a limited amount of the two fluorescent proteins in extremely pure form, a large number of fishes were required. The result of 60 steps of purification on material from 50,000 jellyfish was just 200 mg of aequorin, one of the fluorescent proteins. Finally, two laboratories, one in Georgia and the other in Japan independently cloned the gene for aequorin in 1986. This allowed them to start experimenting with genetic modification on it. The green fluorescent protein (GFP), the second of the jellyfish fluorescent proteins, was cloned in 1992 by Douglas Prasher and colleagues at the Woods Oceanographic Institution in Massachusetts (Prasher et al., 1992)which is comprised of modified amino acid (aa. Blue light from aequorin was absorbed by GFP, which then fluoresced green. Chalfie's lab inserted the gene into bacteria and frog cells after receiving samples of it from Prasher. Chalfie and Prasher reported success. The viability of this had not been made obvious. Chalfie had been worried that more jellyfish molecules might be required in order to construct or activate the molecule. Although it wasn't the case, there were still issues. GFP required to be activated by the laser in a microscope to be seen in a cell, but it was not bright enough to provide a distinct signal. Additionally, it did not function well at body temperature, making it challenging for researchers to introduce it into mammalian cells.

Rebuilding the protein may be the answer to these issues, but in order to know what changes to make, a detailed analysis of the protein's structure was required. The GFP molecule was crystallized by S. James Remington, a structural biologist at the University of Oregon, who then utilized X-rays to create a high-resolution map of the molecule. His team observed that GFP has a barrel-shaped domain made up of 11 beta strands and one alpha helix, a structure that is distinct from other proteins. The three amino acids of the helix are involved in both absorbing and emitting light. One of Remington's partners, Roger Tsien, and his colleagues at the University of California were inspired by this structure to find ways to enhance the protein. The wavelength of light required to activate the protein was adjusted by changing one of the three amino acids, enabling the protein to be activated by common lasers used in microscopes. Additional adjustments increased the module's brightness and enabled GFP to function effectively at body temperature. The molecule has been modified by Tsien's lab and others since the late 1990s so that it can be activated in various ways and emit various colors, such as blue, cyan (green-blue), and yellow. Proteins from coral and other animals have been used to create more fluorescent tools. The proteins are now thousands of times more brilliant than their natural counterparts because of genetic editing. Shimomura, Chalfie, and Tsien received the 2008 Nobel Prize in Chemistry for their enormous contribution to biology (Weiss, 2008). GFP's fluorescent module can be linked to other genes, which frequently causes cells to produce fluorescence. If a laser microscope is used to illuminate the changed protein, it will emit light. This is useful for a number of things. The first step is to determine whether specific cell types can produce a certain chemical. Finding out where a protein functions in the cell is another use for proteins.

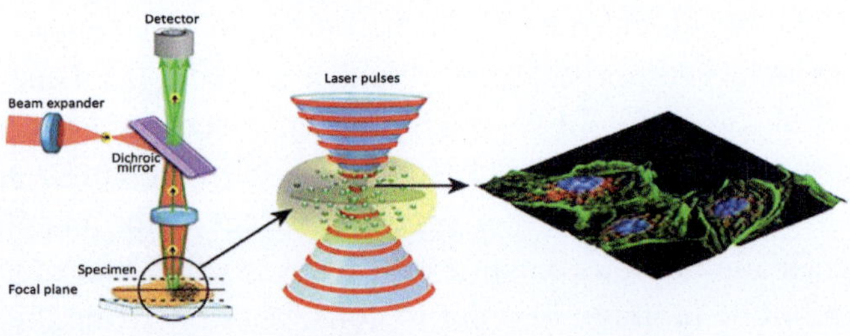

Figure 1.16. Fluorescence microscopy. (Source: Ishikawa-Ankerhold HC, Ankerhold R, Drummen GPC. Advanced Fluorescence Microscopy Techniques—FRAP, FLIP, FLAP, FRET and FLIM. *Molecules*. 2012; 17(4):4047-4132. https://doi.org/10.3390/molecules17044047, Creative Commons License).

Finding out whether two proteins cooperate, as, for instance, when a protein is required to activate another gene, is another application for fluorescence microscopy (Figure 1.16). You may check this by labeling the two molecules with various colors and keeping an eye out to see if the production of one protein is followed by the appearance of the other. The science of how GFP and related proteins collect energy from one form of light and then emit it as another is the basis for some of the more complex applications of these proteins. Each protein can only perform this function once. This means that activating a molecule with the microscope's laser causes it to become illuminated. This finding has been used to develop a technique for researching how the cell produces, transports, and recycles proteins. The laser can be pointed at a membrane and left there until a certain molecule has been bleached, allowing researchers to determine how frequently a cell reloads its membrane with a specific protein, for example. The cell is actively producing new copies of the fluorescent molecule and delivering them to the membrane if a fluorescent form of the molecule reappears there. Researchers were able to take their first look at how frequently the cell produces and replaces receptors due to the technique known as fluorescence recovery after photobleaching (FRAP) (Meyvis, De Smedt, Van Oostveldt, & Demeester, 1999). FRAP has been utilized by Philippe Bastiaens to show that some receptor protein types are recycled. When a protein's activity changes, such as when a signaling molecule has been chemically activated, the signal changes. This makes it possible to conduct a variety of additional studies. For instance, scientists can determine if two molecules attach to one another, which is crucial when evaluating prospective new drugs. The majority of drugs are microscopic compounds that attach to crucial proteins and alter their functions. A fluorescent marker may be used to label the target protein. Scientists can put the drug candidate into the cell and then keep an eye out for changes in the molecular fluorescence. This can reveal whether the drug has bound and whether this has altered the protein's activity.

A method known as fluorescence resonance energy transfer (FRET) can be used to determine how and where two molecules that have been labeled with different tags attach to one another (Selvin, 1995). The fluorescent modules of the two proteins become quite near to one another during binding. Some of the energy that one of them emits when activated by a laser may be absorbed by the partner molecule. In a similar experiment, Bastiaens and his coworkers demonstrated via FRET that the $ErbB_1$ receptor, which receives signals from proteins like the epidermal growth factor, can send a signal into the cell very quickly. Multiple receptors must often be activated in order to elicit a robust response from the cell since receptors might occasionally be accidentally activated. However, it was demonstrated that EGF may elicit a potent response even if just one receptor is excited; the stimulated receptor produces a wave of transmission that activates more copies of $ErbB_1$ in the membrane. None of the techniques that can achieve better resolution, like X-ray crystallography, can see molecules in their actual, natural environments. It will always be important to examine the fundamental components of life in the context of living cells. The resolution required to see individual molecules is not present in even the most potent electron microscopes. However, they can get a peek/look of massive proteins or protein machines. That might at least enable researchers to identify specific cell machinery in a cell image by looking at it. Data from Gavin and her coworkers have been used by bioinformaticians like Rob Russell of the European Molecular Biology Laboratory in Heidelberg to attempt and find a solution to the issue. In order to develop a reasonable assumption about the general morphologies of molecules, they use structural data from individual proteins in yeast cells and attempt to fit the molecules together by computer analysis. High-resolution microscope pictures could be interpreted. Novel approaches to cell exploration are being carried out using electron microscopes.

Cryo-electron tomography is now providing precise maps of the interior of the cell that will be very helpful when looking for protein complexes. Researchers in Germany developed the electron microscope in the 1930s. Since electrons have a shorter wavelength than photons, their resolution is 1,000 times higher than that of a light microscope, and some electron microscopes have a 2,000,000x magnification capacity. Because even air scatters electrons and would produce an image with excessive noise, samples must be viewed in a vacuum for the instrument to function. Therefore, it is impossible to study living cells or tissues. Another issue is that the sample gradually degrades after being exposed to the electron beam. The samples have been coated in gold or other metals, frozen, and other techniques have been developed to decrease noise and lengthen their lifetime. The transmission electron microscope (TEM) and scanning electron microscopes (SEM) are two examples of such instruments now in use. By directing an electron beam towards a sample and catching the electrons that pass through, the TEM operates. Low-energy electrons that have been stimulated by the beam are captured by the SEM. Eliminating noise is one technique to improve the detail of the images, and numerous laboratories started employing software to do this in the 1990s. The plan was to superimpose hundreds or thousands of pictures of the same kind of thing. Since noise was random, the software searched for recurring features in the photos and then removed them. The technique required a large number of photos that not only featured the same object but also displayed it in the exact same position. The icosahedral viruses were the subject of analysis. These contagious organisms are named after their 20-sided structure, which resembles a soccer ball.

These viruses' surfaces are highly symmetrical, which means that no matter which way they are rotated, they essentially look the same. Since an infected cell frequently harbors hundreds of copies of these viruses, many samples may be found there. With a resolution greater than nine angstroms, this technique has produced images of viruses like the herpes simplex virus-1. Taking many images of the same thing from various perspectives and combining them is a way to get a clearer, sharper view of the objects within cells and to perceive them in three dimensions. The basis for cryo-electron tomography is this. Tomography is a technique that uses a variety of waves, such as X-rays and ultrasound, to focus on a specific depth inside an object, similar to how a camera lens may be adjusted so that objects in the middle are sharp while objects closer to or farther away are indistinct (Duric et al., 2005). Ultrasound tomography can show doctors an image of an embryo inside the mother's womb since the waves penetrate the surface, and cryo-electron tomography can show them objects inside cells. The sample is positioned in the electron microscope and rotated such that the electron beam enters it from various angles. The pictures are then stitched together to create a single, three-dimensional image. The technique has been employed by Stephen Fuller's lab at the University of Oxford in Great Britain to investigate how cells produce HIV. HIV infects cells by hacking its genetic material into the DNA of the host cell. Cells start making the building blocks required to create new copies of the virus as soon as that information becomes active. The components must be brought together, and then filled with fresh copies of the viral genetic material. The components are created in different parts of the cell. The majority of the copies of HIV produced by an infectious cell are incorrectly constructed and no longer contagious since this procedure must be carried out so precisely. However, only a few must be successful in doing so in order to invade other cells, eventually compromising the host's health and moving on to new victims.

In order to distinguish between the creation of infectious and harmless copies of HIV, Fuller and his colleagues have been carefully examining the method by which it is created. Cryo-electron tomography was employed by Fuller's team in 2006 to follow the development of the viral core, a cone-shaped intracellular compartment containing HIV RNA and proteins. The membrane-covered sphere that will serve as the virus's outer capsule is where the cone's proteins are assembled. The cone differs in size and shape from other elements of HIV; Fuller was interested in how the cell could tolerate these variations and still manage to create an effective, correctly shaped, sealed compartment. His research team found that the cone forms at the narrow end and expands outward. Once it has finally reached the membrane, the ends will bend and fuse together to seal the compartment. Another study done in 2006 by Dirk Schuler's lab at Ludwig Maximilian University in Munich utilizing cryo-electron tomography discovered a strange new structure in magnetotactic bacteria. The waters are home to a large number of these organisms. The magnetosome, a microscopic organ inside of them that stores crystals of the mineral magnetite, is what makes them special. The magnetic properties of the crystals aid in providing the bacteria with a sense of direction. The cells move toward tiny pockets of oxygen on the ocean floor using the magnetic field of the Earth. Cryo-electron tomography was used by Schuler and his colleagues to study the bacteria in an effort to learn how the cell managed the placement of magnetosomes (Katzmann et al., 2011). The organelles are kept in a straight line within the cell, despite the fact that they should form clumps due to their magnetic attraction to one another. The protein-based structure that held them in place was visible in the microscope pictures as a new, fiber-like structure. The magnetosomes appear to be glued to certain locations along the fiber by another protein in the cell.

1.11 NUCLEIC ACIDS

Researchers have learned a lot about the structure and activity of individual molecules since the creation of molecular biology in the 1940s and 1950s. The internal architecture of protein machines should be discovered, and their dynamic interactions in the cell should be studied. A cell that may have existed three billion years ago is the ancestor of every living cell today. This organism is referred to as the last universal common ancestor. Bacteria, archaea, and eukaryotes are the three major branches of life that have descended from this organism. The final category consists of all plants, all animals, and a broad variety of single-celled organisms, including yeast. Because of shared information contained in their genomes, scientists can conclude that all of these organisms descended from a single ancestor. Each cell is a live representation of the knowledge it got from its parents and their most distant relatives, despite the many modifications and advances that have given rise to at least hundreds of millions of different species. Life on Earth requires the capacity to preserve and safeguard genetic information. DNA serves as the main repository of information. Because they serve as blueprints for the production of proteins, messenger RNAs (mRNAs), which are encoded in genes, are transient information storage systems. Each type of cell has a basic set of molecules that it must produce in order to take on a certain form, divide at the appropriate time, and mature properly. The population of proteins and the set of mRNAs, on the other hand, are constantly changing. This kind of cellular memory is very crucial.

While DNA, RNA, and proteins are the building blocks of all life on Earth, there are some significant variances in how very diverse sorts of organisms handle these molecules. Proteins are typically produced by bacteria and archaea. In 1951, when the young American James Watson arrived in Cambridge, England, many scientists had finally come to the conclusion that the genetic code was contained in the DNA molecule. More than 80 years had passed since the molecule was first isolated from pus in the bandages of injured soldiers by a German scientist by the name of Friedrich Miescher. Frederick Griffith, a medical officer in London, conducted experiments in 1928 that demonstrated how molecules may be transferred across different bacteria to give one strain the traits of another. The development of antibiotics rendered Oswald Avery's (1877–1955) nearly 35 years of work on a pneumonia vaccine at the Rockefeller Institute for Medical Research in New York obsolete (Méthot, 2016). He continued where Griffith had left off in 1940 and proved that bacterial DNA carried the transforming information. However, no one knew how DNA could transfer information from one cell to another or whether this also applied to other organisms. Answers might be found through comprehending the genetic material's chemical and physical makeup, according to scientists. William Astbury, a physicist at the University of Leeds in England with a focus on textiles, began studying DNA in 1937. He had been utilizing X-rays to determine the diffraction patterns of wool and other fibers when he discovered that DNA fibers had a consistent helical structure, similar to a spiral staircase. Adenine, Cytosine, Guanine, and Thymine, the four chemical constituents of DNA, had to combine with molecules of phosphate and sugar to form this structure.

Chapter

2 PRINCIPLES OF GENETICS

CONTENTS

2.1 INTRODUCTION

Any biological species has members that share certain traits but differ in others. For instance, the observable qualities that characterize humans as a species are shared by all people. We are classified as a type of vertebrate since we have a backbone and spinal cord. As a type of mammal, we are warm-blooded and provide milk from our mammary glands for the nourishment of our young. More specifically, we are a species of monkey that typically stands up straight, has long legs, a little amount of body hair, a huge brain, a flat face with a prominent nose, protruding chin, unique lips, and small teeth. These characteristics distinguish us from other primates like chimpanzees and gorillas (Figure 2.1).

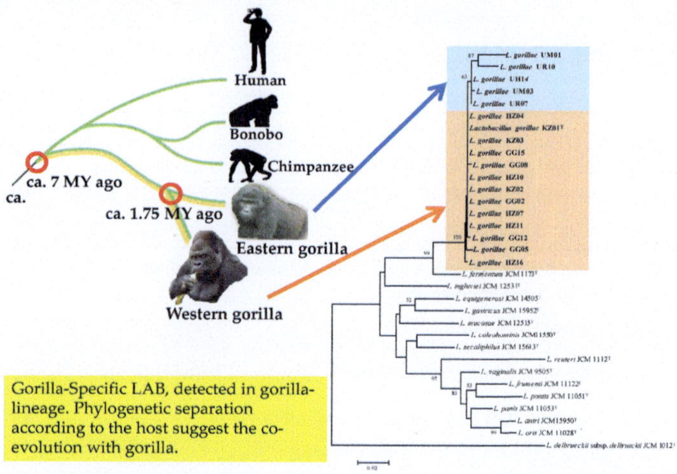

Figure 2.1. Gorilla. (Source: Tsuchida S, Kakooza S, Mbehang Nguema PP, Wampande EM, Ushida K. Characteristics of Gorilla-Specific *Lactobacillus* Isolated from Captive and Wild Gorillas. *Microorganisms*. 2018; 6(3):86. https://doi.org/10.3390/microorganisms6030086, Creative Commons License).

The biological traits that make us who we are as a species are inherited, but they are constant from one individual to the next. However, there is also a great deal of diversity within the human species. The differences between individuals in traits like height, weight, eye color, skin color, and personality traits are enormous. Health-related characteristics such as a propensity for high blood pressure, diabetes, substance abuse, mental depression, and Alzheimer's disease might also vary. Some of these characteristics are passed down biologically, while others are passed down culturally. The native language we speak is inherited culturally, whereas eye color is inherited biologically. Environmental and biological factors work together to determine many traits. Genetics is the study of biologically inherited qualities; for instance, weight is controlled in part by genetics, but also in part by eating habits and extent of physical activity (Pickrell et al., 2016). Environmentally influenced traits are one of the qualities that are examined in genetics. The underlying idea behind genetics is that genes, which are components of heredity that are passed from parents to children during reproduction, are what determine inherited features.

Figure 2.2. Gregor Mendel: Woodcut by Michael Florian issued at the centennial of Genetics Laws. Text in Czech states: 100 years of Genetic Laws by Gregor Johann Mendel, Abbot of Augustinian Monastery in Old Town of Brno. (Source: Vyskot B and Siroky J (2022) Bicentennial of Gregor Johann Mendel's birth: Mendel's work still addresses geneticists in 2022. Front. Plant Sci. 13:969745. doi: 10.3389/fpls.2022.969745, Creative Commons License).

2.2 MENDEL EXPERIMENTS

Gregor Mendel (Figure 2.2) conducted research with garden peas to discover the components of heredity and some fundamental principles guiding their transmission from generation to generation. In 1866, his findings were published. Mendel's experiments rank among the best in the history of experimental science in terms of their delicate conception, meticulous execution, and eloquent interpretation. Mendel explained his findings in terms of a few laws that describe how traits are passed down from parents to children.

Figure 2.3. Friedrich Miescher. (Source: By copied from http://www.pbs.org/wgbh/nova/photo51/images/befo-miescher.jpg, Public Domain, https://commons.wikimedia.org/w/index.php?curid=789048, Creative Commons License).

Friedrich Miescher (Figure 2.3) made a new type of weak acid material, which was prevalent in the white blood cell and salmon sperm nuclei, three years later, in 1869 (Dahm, 2008). He had no way of knowing at the time that it would turn out to be the material that makes up genes. The molecule that makes up a gene, once known as Miescher's weak acid, is now known as deoxyribonucleic acid (DNA). The discovery that the nuclei of male and female reproductive cells merge during fertilization in the 1870s led scientists to realize the significance of the cell nucleus in heredity. This observation suggested that the inherited traits were caused by something within the sperm and egg nuclei. The next important development was the identification of chromosomes, which are thread-like components of the nucleus that can be seen under a light microscope when stained with specific dyes. Chromosomes have a distinctive splitting behavior during cell division, which guarantees that each daughter cell obtains an identical complement of chromosomes. By 1900, it had been evident that while chromosomal counts vary between species, they are stable within each species. It seems plausible that chromosomes were the genetic carriers based on their properties.

By the 1920s, mounting evidence pointed to a close connection between genetic material and DNA. Studies utilizing specific stains demonstrated that chromosomes contain DNA in addition to certain proteins. Investigations have revealed that whereas the quantity and types of proteins and other molecules vary significantly among various cell types, DNA is virtually always present in equal amounts in all cells of a given species. Because incomplete chemical tests of DNA had wrongly concluded that it lacked the chemical variety required for a genetic substance, the indirect evidence that genes are DNA was ignored. Proteins, on the other hand, were recognized to be a remarkably diverse group of molecules. So, on the basis of some information, it was commonly believed that proteins served as the genetic material and that DNA only served as the framework for chromosomes. Experiments that claim to show that DNA is the genetic material are also required to show that proteins are not the genetic material in order to refute the widely held belief that genes are proteins.

2.3 BACTERIAL STUDIES

Studies on the bacteria that cause pneumonia provided the first proof that genes are DNA. *Streptococcus pneumoniae* strains with the ability to produce a slimy capsule around each cell are the ones that cause bacterial pneumonia in mammals. Capsule-lacking strains do not cause pneumonia. The bacterium is shielded from the immunological response of the infected animal by the polysaccharide found in the capsule, which also allows the bacteria to cause disease. A colony is a visible collection of cells that forms when a bacterial cell is grown on solid medium after going through numerous cell divisions. The colony appears shiny or smooth (S) due to the encircling capsule. Some *S. pneumoniae* strains are incapable of producing capsules (Yother et al., 1986). They consequently produce colonies with a rough (R) surface. Because the host immune system attacks the bacteria when the R strains are absent from the body, they do not cause pneumonia. The progeny produced by cell division of both types of bacteria breed true in the sense that they have the parent's capsular type, either S or R. When mice are injected with either living R cells or dead S cells that have been heated to kill them, the mice don't get sick. However, it was found in 1928 that injecting mice with a mixture containing a large number of dead S cells and a small number of surviving R cells frequently resulted in pneumonia. Even though the injected S cells had been destroyed by heat, bacteria isolated from blood samples of the infected animals developed S cultures with a capsule typical of the injected S cells. Therefore, the chemical that was included in the substance containing the dead S cells that was injected must have been able to change, or transform, otherwise benign R bacterial strain cells into S strain cells with the ability to fight the mouse immune system, multiply, and cause pneumonia. In other words, an R cell underwent a genetic metamorphosis into an S cell. The descendants of the altered bacteria carried the new genetic traits.

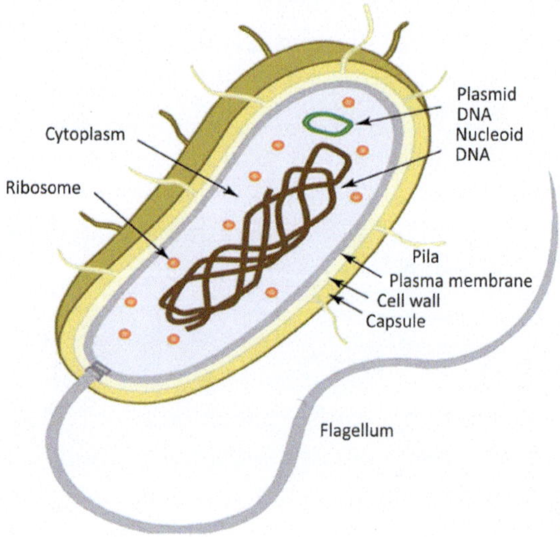

Figure 2.4. *Escherichia coli*. (Source: Basavaraju, M., & Gunashree, B. S. (2023). *Escherichia coli*: An Overview of Main Characteristics. IntechOpen. doi: 10.5772/intechopen.105508, Creative Commons License).

Another significant discovery involves a virus that affects bacterial cells. Because it infects bacterial cells, the virus, T2, is sometimes referred to as a bacteriophage or phage (Dabrowska et al., 2007)non-antibacterial effects of bacteriophage T4 in mammals (binding to cancer cells in vitro and attenuating tumour growth and metastases in vivo. *Escherichia coli*, (Figure 2.4) an intestinal bacteria, is infected by T2. Its head, which houses the phage DNA, collar, tail, and tail fibers make up its extremely small but complicated structure. The first step of a T2 infection is the attachment of a phage particle to the bacterial cell wall via its tail tip. Next, phage material is introduced into the cell and multiplied to create one hundred or more progeny phages, which are then released by disrupting the bacterial host cell. Because proteins often include some sulfur but no phosphorus and DNA typically contains neither, it is possible to distinguish between the DNA and proteins in a phage particle using radioactive isotopes of each element. Alfred Hershey and Martha Chase, who were both working at the Cold Spring Harbor Laboratories in 1952, made use of this distinction. By then, it was already known that T2 particles are made up roughly equally of DNA and protein. By infecting *E. coli* cells that had been raised for several generations in a medium containing ^{32}P (a radioactive isotope of phosphorus), Hershey and Chase were able to create viral particles carrying radioactive

DNA. They then collected the phage progeny. Using a medium containing ^{35}S, more particles with tagged proteins were produced in the same manner. To follow the proteins and DNA independently, nonradioactive *E. coli* cells were infected with phage tagged with either ^{32}P or ^{35}S. Centrifugation was used to concentrate the infected cells, which were then resuspended in fresh medium and mixed in a blender to remove any associated phage material. The subsequent course of the infection was shown to be unaffected by the blending, which suggests that the genetic material must enter the infected cells relatively quickly after phage attachment. Most of the radioactivity from the ^{32}P-labeled phage was discovered to be associated with the intact bacteria when they were separated from the material removed by blending; however, when the infecting phage was labeled with 35S, only about 20% of the radioactivity was associated with the bacterial cells. These findings showed that a T2 phage mainly delivers its DNA but just a small amount of its protein to the cell it infects.

The crucial discovery was that the offspring phage particles inherited roughly 50% of the transferred ^{32}P-labeled DNA but less than 1% of the transferred ^{35}S-labeled protein. The Hershey-Chase experiment was not nearly as stringent as the transformation studies in implicating DNA as the genetic material because some protein was delivered to infected cells and transmitted to the offspring phage. Nevertheless, the experiment had a significant impact because it was consistent with the DNA hypothesis. Today, the equivalent of the transformation experiment is routinely performed at many research facilities across the globe using cells from animals, plants, yeast, bacteria, or yeast maintained in culture. These studies demonstrate that DNA, as well as phage T2, is the genetic material in these species. The generalization that DNA is the genetic substance in all cellular organisms has no known exceptions. However, it is important to keep in mind that the genetic material of some viruses is composed of RNA, a different kind of nucleic acid. Even after learning that genes are made of DNA, many unanswered issues persisted.

In an effort to find answers to these issues, several academics started delving into the intricate chemical structure of DNA in the early 1950s. James Watson and Francis Crick suggested the first and essentially true three-dimensional structure of the DNA molecule in 1953. DNA is made up of two lengthy chains of subunits that are wrapped around one another to form a double-stranded helix in the Watson-Crick structure. The double helix is right-handed, thus as one moves down the strand, each chain advances in a clockwise direction (Conner et al., 1982). If you imagine yourself gazing up into the structure from the bottom, you can see the right-handed coiling. Each individual strand's backbone coils in a clockwise manner. Each strand's nucleotide subunits, also known as bases, can include any one of four different chemical components.

2.4 RESEARCH STUDIES ON DNA

This theory explains how DNA's limited number of nucleotides can encode the vast quantity of information required to create an organism. The genetic information is encoded by the bases that make up DNA, and the base sequence is fully uncontrolled. The Watson-Crick pairing is another name for the complementary pairing (Gao et al., 2004). The base pairs are shown as the spheres that occupy the double helix's interior in the three-dimensional structure. The base pairs are arranged in a nearly flat stack perpendicular to the long axis of the double helix, resembling a roll of coins. Biologists typically refer to a DNA molecule's separate strands as single-stranded DNA and its double helix as double-stranded DNA or duplex DNA when discussing it. Each DNA strand has a polarity, or directionality. The angle that the nucleotides are pointing at determines the polarity. The tail end of the strand is referred to as the 3' end, while the starting end is referred to as the 5' end. The 5' end of one strand is aligned with the 3' end of the other in double-stranded DNA, which has paired strands that are oriented in opposite directions. The original (parent) duplex strands split apart during replication, and each one acts as a pattern or template for the creation of a new strand (replica). Each base in the replica is complementary (Watson-Crick pairing) to the base across the way in the template strand when the replica strands are created by adding successive nucleotides. It is actually a complicated process involving numerous enzymes and other proteins to solve its chemical and geometrical issues. It is crucial to remember that a duplex molecule replicates into two daughter molecules, each of which has a nucleotide sequence that is exactly like the parent strand.

It was already evident at the beginning of the 20th century that proteins were involved in the majority of cellular metabolism. It was previously understood that proteins were necessary for the breakdown of organic molecules in order to produce the chemical energy required for cellular activity. It was understood that they were necessary for the fusion of simple molecules into more intricate molecules and biological structures. The word enzyme was first used in 1878 to describe the biological catalysts that quicken metabolic reactions in organisms. By 1900, it had been established that enzymes were proteins, largely due to the research of German scientist Emil Fischer. Other proteins are essential parts of cells; for instance, structural proteins give cells their shape and mobility, other proteins create pores in cell membranes and regulate the flow of small molecules into and out of cells, and other proteins control cellular activities in response to molecular signals from the outside environment or from other cells. The idea put out by Watson and Crick that genetic information in DNA is encoded in a way similar to letters (the bases) printed on a strip of paper is entirely correct. However, it required another 20 years of work to understand the specifics of the genetic code and how it is decoded. The extensive body of research revealed that the genetic code for a protein is located in a DNA strand within a region of DNA that regulates protein synthesis. Each succeeding word in the DNA strand specifies the following chemical subunit to be added to the protein as it is being produced, and the genetic information encoded in this strand is decoded in a linear order. In the genetic code, each word is made up of three nearby bases.

For instance, the DNA base sequence ATG designates the amino acid methionine (Met), TTT designates the amino acid phenylalanine (Phe), GGA designates the amino acid glycine (Gly), and GTG designates the amino acid valine (Val). The fundamental idea of molecular genetics is that proteins are made up of RNA and DNA, respectively. The core dogma's key idea is that ribonucleic acid, an intermediate molecule, rather than DNA itself, directly codes for proteins. RNA and DNA have a similar but distinct structural composition. Rather than being deoxyribose, the sugar is ribose. The most common form of RNA is single-stranded (not duplex), and it contains uracil (U), a nucleotide that replaces thymine (T) in DNA. There are actually three distinct RNA types that take part in and contribute to the production of proteins. It is simple to understand how genetic information is transmitted from DNA to RNA. One of the DNA strands is employed as a template to create a complementary strand of RNA as soon as the DNA splits open. Transcription is the process of creating an RNA strand from a DNA template, and the resulting RNA molecule is the transcript (Alberts et al., 2002). With the exception of U (which pairs with A) being present in the RNA in place of T, the base sequence in the RNA is complementary to the DNA template. An RNA strand has a 5' end and a 3' end that are both defined by the orientation of the nucleotides, just like DNA. In the RNA—DNA duplex created during transcription; the polarity of the RNA strand is the opposite of that of the DNA strand. The 5' end of the RNA transcript is initially synthesized. Translation is the process of creating a protein under the control of an mRNA molecule. The molecules that actually carry out the translation are the tRNA molecules, even if the sequence of mRNA bases codes for the sequence of amino acids. Codons, which are collections of three bases, are used to transcribe the mRNA molecule. There is a tRNA molecule that has a complementary group of three adjacent nucleotides that can couple with each codon in the mRNA. The incomplete protein chain is joined to the amino acid on the tRNA at each stage of protein synthesis, lengthening the protein chain by one amino acid, when the appropriate tRNA with an attached amino acid comes with mRNA.

2.5 AMINO ACIDS

The chain of amino acids created during translation is technically referred to as a polypeptide. A protein can be made up of many polypeptide chains that assemble after translation, which is how it differs from a polypeptide. Some proteins (encoded by the same gene) consist of two or more identical polypeptide chains, while others are made up of two or more distinct polypeptide chains (encoded in different genes). Hemoglobin is made up of four polypeptide chains that are encoded by two distinct genes. The beta-globin gene produces two of the chains, which are beta polypeptide chains, while the alpha-globin gene produces the other two chains, which are polypeptide chains. Any heritable gene change is referred to as a mutation. A mutation is a change in the order of the bases along the DNA, according to the Watson-Crick model of DNA. Simple modifications include switching out one pair of bases in a duplex molecule for another pair of bases. For instance, in a duplex molecule, A—T pair could change into T—A, C—G, or G—C. More complicated changes to the base sequence, like the deletion or insertion of base pairs, are also possible. One of the paradoxes of genetics is that, while the relationship between genes and proteins is conceptually direct, the relationship between genes and phenotypes is unquestionably complex. This is a result of the biochemical complexity of organisms. The majority of an organism's observable characteristics are the end product of many genes working together in conjunction with environmental conditions. A mutant gene may have an impact on a variety of features that seem unconnected. The multiple forms of the mutation are referred to as pleiotropic effects, and the mutation is said to demonstrate pleiotropy as a result. The primary oxygen-carrying protein of red blood cells, which is affected in sickle-cell anemia, is an example of a mutation in humans that has many pleiotropic effects (Randolph & Zhao, 2015).

The gene encoding beta-globin, which codes for the beta polypeptide chains found in the oxygen-carrying protein of red blood cells, is the site of the underlying mutation in sickle-cell anemia. In place of the typical A—T base pair, the mutant form of the gene has a T—A base pair. The mutation resulted in a GUG codon in the mRNA. Because GUG specifies valine (Val), this amino acid is substituted for glutamic acid (Glu) at position 6 of the polypeptide. The hemoglobin protein has a propensity to form lengthy, needle-like polymers because of the deficient beta polypeptide chain. When polymerization occurs in red blood cells, the cells are bent into crescent-shaped structures resembling sickles. Some of the abnormal red blood cells are rapidly destroyed, decreasing the blood's ability to carry oxygen and contributing to anemia, while others may group together and obstruct capillary blood flow. Weakness, jaundice, and poor growth are all effects of such anemia. The most common cause of mortality in children with the disease is bacterial infections because affected individuals are often so delicate. Sickle-cell anemia is a severe genetic condition that frequently leads to premature death, but it is rather common in parts of Africa and the Middle East where *Plasmodium falciparum*, (Figure 2.5) a parasitic protozoan, is a common cause of malaria. Malaria and sickle-cell anemia are related.

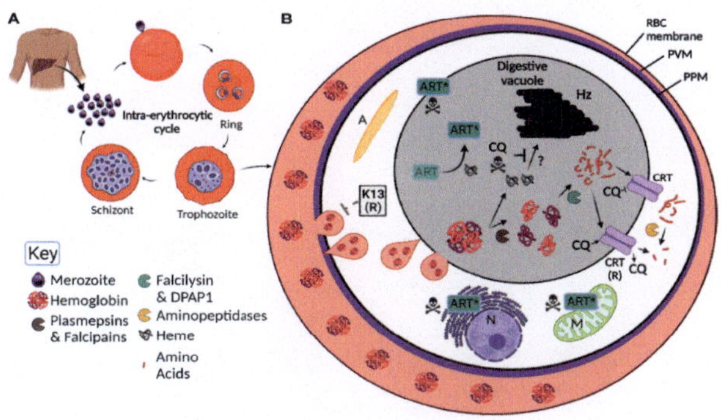

Figure 2.5. The lifecycle of the parasite *Plasmodium falciparum*. (Source: Edgar RCS, Counihan NA, McGowan S and de Koning-Ward TF (2022) Methods Used to Investigate the Plasmodium falciparum Digestive Vacuole. Front. Cell. Infect. Microbiol. 11:829823. doi: 10.3389/fcimb.2021.829823, Creative Commons License).

The parasite spreads from a mosquito to a human during its life cycle by the bite of the insect. The parasite's earliest stages of infection begin in liver cells, where specialized forms are created that can infect and grow in red blood cells. The lethargy, anemia, and jaundice that are symptoms of malaria are brought on by a widespread infection of red blood cells, which reduces the capacity of the blood to carry oxygen. However, it is believed that in people with the mutant beta haemoglobin chain, the infected blood cells go through sickling and are quickly removed from circulation. Thus, the parasite's ability to multiply within red blood cells is restrained, and the severity of the malarial infection is lessened. As a result, the prevalence of the genetic disease sickle-cell anemia and the parasitic disease malaria are genetically balanced. More people die from sickle-cell anemia (Figure 2.6) than are saved by malaria prevention if the mutant beta hemoglobin is too common; on the other hand, fewer people die from sickle-cell anemia but the benefit is negated by more malaria fatalities if the mutant beta hemoglobin is too rare. Each trait requires a large number of genes for correct physiology, metabolism, and development. Consequently, multiple genes may have an impact on a single trait.

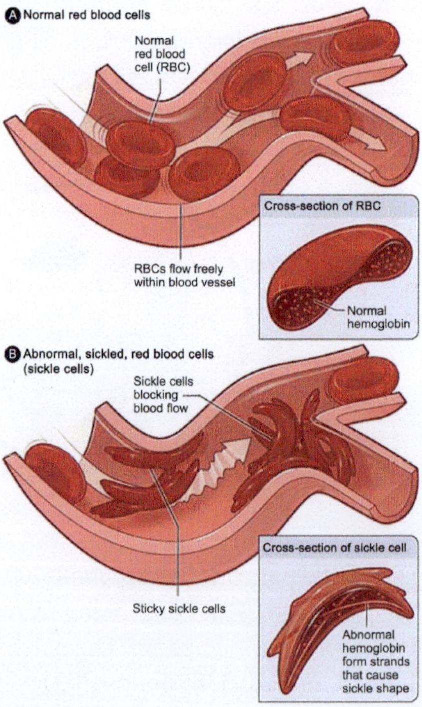

Figure 2.6. Sickle-cell anemia. (Source: By The National Heart, Lung, and Blood Institute (NHLBI) - http://www.nhlbi.nih.gov/health/health-top-

ics/topics/sca/, Public Domain, https://commons.wikimedia.org/w/index.php?curid=19198765, Creative Commons License).

2.6 DROSOPHILA: THE BASIS OF GENETICS

In Drosophila (Figure 2.7), Vermilion (v) and cinnabar are important genes (cn) (Cornel et al., 1997). These genes produce the enzymes V and Cn, which are involved in the metabolic process that turns the amino acid tryptophan into the pigment xanthommatin, which is seen in brown eyes.

Figure 2.7. Drosophila. (Source: By Sanjay Acharya - Own work, CC BY-SA 4.0, https://commons.wikimedia.org/w/index.php?curid=63862733, Creative Commons License).

A distinct enzyme that is expressed by a different gene catalyzes each step in the process. Environment and genes also interact. Consider the condition anemia, which describes a weakness brought on by a deficiency in either red blood cells or blood volume, to understand how genes and environment interact. Anemia comes in a variety of forms. Sickle-cell anemia is one type of anemia that is genetically predisposed. Anemia brought on by persistent dietary iron deficiency and malaria infection are two examples of environmental causes of anemia. Other types of anemia are brought on by a combination of genetic and environmental causes. For instance, eating fava beans can cause severe anemia in people who have a mutant form of the enzyme glucose-6-phosphate dehydrogenase (G6PD), which is crucial for maintaining the integrity of red blood cell membranes (Cappellini & Fiorelli, 2008). This is because the beans contain a substance that causes RBCs to be damaged. The disease is known as favism because of its connection to fava beans. Certain substances like naphthalene, as well as some antibiotics and other medications, can cause red-cell death in patients with G6PD deficiency. Populations in coastal areas around the Mediterranean

Sea have a somewhat high prevalence of G6PD deficiency, which mostly affects males. Red blood cell abnormality is suggested to contribute to an increase in malaria resistance. There are many potential explanations for a complex trait like anemia. Some types are influenced by genes, others by the environment, and yet others need both to express themselves. Anemia serves as an excellent illustration of the genes versus environment debate because several varieties of the condition have already been identified and given genetic, environmental, or a combination of both causes.

Similar to anemia, most complex features are made up of numerous diseases that are grouped together due to their overall resemblance. The common example of this is heart disease. It is common knowledge that genetic risk factors for heart disease are linked to cholesterol and saturated fat metabolism. There have already been discovered certain uncommon disease subtypes with an important genetic component. Environmental risk factors for heart disease include smoking, being overweight, not exercising, and consuming a diet heavy in saturated fats and cholesterol, among others (Maron et al., 1991). Genetic and environmental factors are crucial in defining the overall risk of heart disease in the population as a whole, and some of these factors act synergistically, which means that the risk from two factors combined is higher than would be predicted from the risk of each factor considered separately. Heart disease serves as an illustration of how genetic and environmental factors can have opposing effects. For instance, a person with a family history of heart disease can significantly reduce their risk with careful dietary choices, regular exercise, quitting smoking, and other lifestyle choices. Another example of an environmental intervention that lowers the overall risk of heart disease is taking medication to lower blood pressure. A common illustration of a complicated trait impacted by many genes as well as several environmental influences is heart disease. This category includes personality and other behavioral traits, which make up the majority of humans. Some qualities are more strongly influenced by genetic variables than others, and it can be very challenging to separate the several aspects of a trait that may have a common ancestor.

2.7 EVOLUTION

The process of evolution, which involves populations of organisms that are descended from a common ancestor gradually becoming more adapted to their environment and occasionally splitting into new species, is what gives rise to the molecular unity of life. According to evolutionary theory, the unity of essential molecular processes results from inheritance from a remote common ancestor that already had a number of mechanisms in place. A type of ribosomal RNA molecule called rRNA present in all of these organisms, with comparable nucleotide sequences, served as the basis for the evolution tree. Most bacteria and cyanobacteria (formerly called blue-green algae) (Figure 2.8) are included in the group microbes.

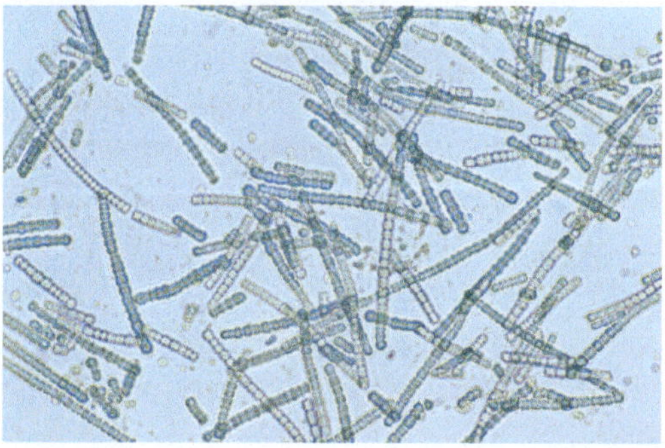

Figure 2.8. Cyanobacteria. (Source: https://en.wikipedia.org/wiki/File:CSIRO_ScienceImage_4203_A_bluegreen_algae_species_Cylindrospermum_sp_under_magnification.jpg, Creative Commons License).

These cyanobacterial cells are surrounded by a cell wall, divide by binary fission, and lack membrane-bounded organelles like mitochondria and nuclei. The Archaea group of microorganisms was initially identified as methane-producing organisms or species that thrive in harsh conditions like hot springs or areas with high salt concentrations, but they are also extensively found in more typical settings. The cells of Archaea do not have interior membranes, unlike bacteria. According to DNA sequence research, the mechanisms for DNA replication and transcription match those found in Eukarya, whereas the metabolism of Archaea is most similar to that of bacteria. However, only around 50% of their genes are exclusive to Archaea. Eukarya species have cells that have a complex internal membrane network, a membrane-bounded nucleus, and mitochondria. Their DNA is structured into real chromosomes, and mitosis is used for cell division. Plants, mammals, fungi, and several single-celled species, such as amoebae and ciliated protozoa, are all considered to be eukaryotes. Prokaryotes, which literally means "before nucleus," is the term used to describe a bigger assemblage that includes both Bacteria and Archaea.

Since differences between species are inherited, mutation must have caused the original differences. To understand why organisms are adapted to live in their surroundings, such as why marine mammals have unique adaptations that allow for swimming and diving, or why desert mammals have unique adaptations that allow them to survive with little to no water, mutations alone are insufficient. A mutation is a chance event that is not intended to achieve a specific adaptation purpose, such as longer fur in Arctic mammal species. Natural selection, according to Darwin, is the process by which organisms that have specific mutations or combinations of mutations that enable them to survive or reproduce more effectively in the current environment will leave more offspring than other organisms and thus contribute their favorable genes disproportionately to future generations (Ayala, 2007). The entire species undergoes genetic transformation if this process is repeated over the course of many generations because an ever-increasing part of the population acquires the advantageous mutations. Without a doubt, the main mechanism resulting in the genetic adaptation of organisms to their environments is natural selection. The adaptive argument, for instance, claims that the color of blood is red because it frightens opponents when a person is harmed. This theory almost makes sense, but in reality, oxidized iron is what gives blood its red color, much like rust does. An iron atom is carried by each hemoglobin chain, and oxidized iron is red due to physics, not biological evolution. For instance, certain features exist because they are pleiotropic effects of genes chosen for other purposes rather than because they are selectively advantageous in and of themselves. Some significant occurrences in the history of life may also be largely the result of chance. Many evolutionary biologists now think that a global extinction was caused 65.3 million years ago when an asteroid collided with the Pacific Ocean off the Yucatan Peninsula and sent so much debris into the atmosphere that Earth became completely dark for years. Even while the mass extinction brought on by this catastrophe was not the biggest in Earth's history, it did bring an end to all dinosaur species and nearly 90% of other species. Dinosaurs were a very diversified and adaptable collection of species up until that point.

2.8 GENETIC VARIATIONS

Although organisms of the same species share some properties, they might diverge greatly in a number of other traits. Many variations between organisms are caused by either genetic variations, environmental influences, or both. The study of genetics focuses on inherited qualities, particularly those that are somewhat impacted by the environment. Genes are the components of heredity that are passed down through reproduction from parents to offspring. The chemical foundation of genes was initially discovered by Miescher in the form of a weak acid now known as deoxyribonucleic acid, despite Mendel having first put the sorting of genes in subsequent generations into numerical form. However, it wasn't until the middle of the 20th century that experiments showed that DNA is the genetic material. The investigations of Avery, MacLeod, and McCarty demonstrated that the genetic makeup of bacteria could be changed from one type to another by treatment with purified DNA, providing the first compelling proof of the involvement of DNA in heredity (Lederberg, 1994). In research on *Streptococcus pneumoniae*, pure DNA from disease-causing types was used to turn mutant cells that were unable to cause pneumonia into cells that could. The Hershey-Chase experiment, which demonstrated that the T2 bacterial virus largely injects DNA into the host bacterium (*Escherichia coli*), and that the progeny phages have a significantly higher percentage of parental DNA than parental protein, was a key piece of evidence.

The three-dimensional structure of DNA, which Watson and Crick hypothesized in 1953, provided numerous hints regarding how DNA acts as the genetic material. A DNA molecule is made up of two extended chains of nucleotide subunits that are twisted into a right-handed helix. A (adenine), T (thymine), G (guanine), or C (cytosine) are the four bases that can be found in each nucleotide subunit. The nucleotides in a DNA molecule are paired in its two strands. Every A on one strand is paired with a T on the other, and every G on one strand is paired with a C on the other. The term base pairing refers to the complementary base sequences that run the length of the two paired strands in a DNA duplex molecule. The DNA molecule's structure implied that the bases' order could be used to encode genetic information. Changes in the sequence of bases, such as the substitution of one nucleotide for another or the insertion or deletion of one or more nucleotides, can cause mutations. A form of replication in which the two strands of the parental DNA molecule separate and each strand acts as a template for the creation of a new complementary strand was also suggested by the structure of DNA. The majority of genes produce proteins through coding. A number of different types of RNA are involved in the multistep process that converts genetic information from DNA into protein. An RNA strand structurally resembles a DNA strand, with the exception that the backbone of RNA contains the base uracil (U) rather than thymine and the backbone of DNA contains ribose instead of deoxyribose. RNA is typically found in cells as single, unpaired strands. Transcription, the first stage of gene expression, involves the synthesis of an RNA molecule whose base sequence complements the DNA strand being transcribed. The base sequence in the RNA transcript is translated into groups of three adjacent bases during the process of polypeptide synthesis, which occurs on a ribosome. Different transfer RNA (tRNA) types use base pairing to recognize the codons. Each type of tRNA (Figure 2.9) is linked to a specific amino acid, and when a tRNA base pairs with the correct codon on the ribosome, the amino acid on the tRNA is linked to the growing end of the polypeptide chain.

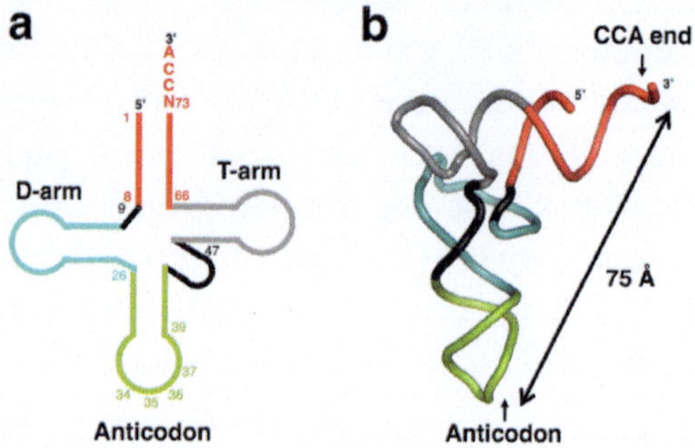

Figure 2.9. tRNA and its structure. (Source: Tamura K. Origins and Early Evolution of the tRNA Molecule. *Life*. 2015; 5(4):1687-1699. https://doi.org/10.3390/life5041687, Creative Commons License).

The start and stop of polypeptide synthesis are designated by special codons. The earliest forms of life used RNA for both genetic information and enzyme catalysis, which is the most likely explanation for why different types of RNA are an integral part of transcription and translation.

2.9 INBORN ERRORS OF METABOLISM

Frequently, the altered protein is functionally flawed, leading to a metabolic inborn error. Because metabolism is made up of a complex branching network of biochemical pathways, the specific way that a metabolic inborn error expresses itself can be very complicated. The majority of an organism's observable traits are the result of numerous genes working together with environmental factors. Because every gene has the potential to affect multiple traits, every trait has the potential to be influenced by multiple genes, and many traits are significantly influenced by both genetic and environmental factors, the relationship between genes and traits is frequently complex. Unrecognized subgroups of many complex features exist that have different genetic or environmental causes. Finding methods to distinguish between the subtypes has frequently led to advancements in genetics. Because all living things share many metabolic and genetic properties, they are all connected. Mutation is the primary cause of organism diversity. Natural selection, on the other hand, is the mechanism through which mutations that are beneficial for survival and reproduction are kept and deleterious mutations are eradicated (Barton & Partridge, 2000). Therefore, the basic method by which organisms get progressively adapted to their environments is natural selection, which was initially postulated by Charles Darwin.

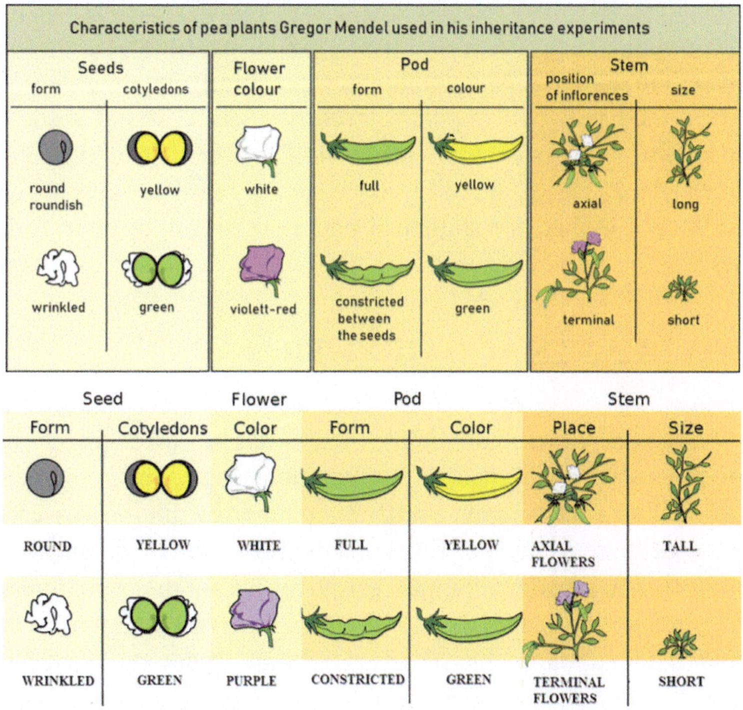

Figure 2.10. Pea plants and Mendel's analysis. (Source: https://commons.wikimedia.org/wiki/File:Mendels_peas.png, Creative Commons License).

Mendel conducted biological studies while living as a monk at the prestigious St. Thomas monastery in the town of Brno (Brünn), in what is now the Czech Republic. Mendel also taught science at a nearby school. The most significant research was done between 1856 and 1863 on sweet pea crosses in a tiny garden plot in a corner of the monastery grounds. In 1866, he published the findings of his research and his analysis in the journal of a local natural history association. He then started correspondence with Carl Nageli, one of the top botanists of the day, in Munich. Nobody, however, recognized the importance of Mendel's work. Mendel's scientific career practically came to a halt when he was chosen as the monastery's head in 1868. Even while contemporary science historians dispute Mendel's motivations for his work, almost all agree that he was an excellent experimenter who conducted meticulous and incredibly well-documented experiments. In his study, he provides the first comprehensive explanation of the statistical principles underlying the transmission of genetic elements from one generation to the next, sometimes known as transmission genetics. Mendel's studies were elegant, which is why they became the cornerstone of genetics. The theories of genetic transmission he deduced from his findings are frequently referred to as Mendelian genetics. Mendel thought quantitatively about features that could be divided into two opposing groups, such as round seeds against wrinkled seeds, which is the main distinction between his methodology and that of other plant hybridizers of his days. He then carried out some crossing experiments and looked for statistical patterns that might point to broader principles.

2.10 PEA PLANTS AND GENETICS

Peas (Figure 2.10) were the plant of choice for Mendel's research for two reasons (Corcos & Monaghan, 1990). First of all, he had access to cultivars that varied in easily distinguishable traits, including round versus wrinkled seeds and yellow versus green seeds. Second, according to his earlier research, peas often reproduce through self-pollination. By opening the petal, which protects the reproductive processes, removing the immature anthers before they release pollen, and dusting the stigma with pollen from a flower on another plant, he was able to cross-pollinate plants to create hybrids. Mendel understood the necessity to investigate inherited traits that varied between pea varieties but were constant within any one variety. For this reason, at the start of his trials, he created true-breeding types in which, when allowed to naturally self-pollinate, the plants only produced offspring that were similar to themselves. The parents for further hybridization came from these several types, which bred for the well-defined characteristics. In order to understand Mendel's procedures and how he interpreted his findings, it is necessary to look at a few of his original experiments. He focused on a pair of traits, such as round versus wrinkled seeds.

All of the hybrid seeds were produced when plants from a variety with round seeds were cross-pollinated with pollen from a variety of plants with wrinkled seeds were round. To distinguish them from the pure-breeding parents, the P1 generation, geneticists refer to the hybrid seeds or plants as belonging to the F1 generation. Mendel also carried out the reciprocal cross, in which the female parents were the plants from the variety with round seeds and the pollen from the variety with wrinkled seeds. The F1 seeds were all round, like they were before. Mendel's crossings between plants that varied in any of the four potential characteristic pairs produced outcomes that were similar. Each time, just one of the parental features was present in every F1 progeny, while the other trait was lacking. Mendel referred to the trait that is expressed in hybrids as the dominant trait and the trait that is not expressed in hybrids as recessive.

Even though hybrid progeny of a monohybrid cross does not express the recessive characteristic, the next generation does when the hybrid progeny is given the opportunity to self-fertilize (Patwardhan, 2022). There were distinct numerical ratios between the two categories. Mendel counted 5474 rounded seeds and 1850 wrinkle seeds. He observed that this ratio was almost three to one. The F2 generation is made up of the offspring seeds created when the F1 generation self-fertilized. Mendel discovered that the proportions of the dominant and recessive phenotypes in the F2 progeny are 3 round: 1 wrinkled. These monohybrid crosses showed that the recessive trait that was not expressed in the F1 hybrids resurfaced in an unmodified form in the F2 generation, varying from the trait present in the initial P1 recessive plants. Mendel said in a letter announcing this discovery that there is nothing to suggest that one of the parents has either inherited or replaced anything from the other, and the two parental features exist, separated and intact, in the F2 generation. Mendel deduced from this result that the qualities in the parental lines were passed down as two distinct elements that retain their purity in the hybrids. To put it another way, the inherited factors do not contaminate one another. Therefore, a plant with the dominant trait may also possess the recessive trait. He made the assumption that each reproductive cell, or gamete, includes one example of each type of plant genetic determinant. He designated the genetic determinant for round seeds as 'A' and that for seeds with wrinkles as 'a'. Mendel hypothesized that all reproductive cells in the true-breeding variety with round seeds would contain 'A', while all reproductive cells in the true-breeding variety with wrinkled seeds would contain 'a'. When the varieties are crossed, the resulting F1 hybrid should inherit one of each of A and a, giving it the genetic composition Aa. The F1 seeds ought to be rounded if A is dominant to a.

The 'A' and 'a' determinants would separate from one another and be present in the gametes in an equal proportion when an F1 plant self-fertilizes. An F2 generation with a genetic makeup of ¼ AA, ½ Aa, and ¼ aa should result from the random pairings of the gametes. The projected ratio of round to wrinkled seeds would be 3:1, with the AA and Aa kinds having round seeds and the aa types having wrinkled seeds. Events that are not mutually exclusive may be independent, which indicates that the possibility of any one event has no bearing on any other events' potential manifestation. It is projected that the ratio of phenotypes in the F2 generation will be 9/16 round yellow, 3/16 round green, 3/16 wrinkled yellow, and 1/16 wrinkled green because the two traits are independent in Mendel's crossings for seed shape and color. These ratios can be calculated by taking each attribute into account independently because they are autonomous. We may anticipate that the F2 generation will contain ¾ round and ¼ wrinkled seeds based just on seed morphology. When only seed color is taken into account, we can anticipate that the F2 generation will be 75% yellow and 25% green. Because the qualities are inherited separately, there should be ¾ of round seeds that are yellow, which means that the overall proportion of round yellow seeds should be 9/16. The proportion of round green seeds that should be present among the ¾ of the seeds that are round 3/16. Similar reasoning can be used to determine how many of the other phenotypic classes there are. When events are independent, the likelihood that they will occur concurrently can be calculated by multiplying.

2.11 GENOTYPES

The genotypes of early progeny have no bearing on the relative proportions of genotypes in later progeny since successive progeny from a cross are also independent occurrences (Hojsgaard, Martínez, & Quarin, 2013)and compete to form a seed as a final goal. This developmental competition was evaluated in tetraploid genotypes of Paspalum malacophyllum in order to understand the low level of sexuality in facultative apomictic populations. Cyto-embryology on ovules, flow cytometry on seeds and progeny tests by DNA fingerprinting were used to measure the relative incidence of each meiotic or apomictic pathway along four different stages of the plant's life cycle, namely the beginning and end of gametogenesis, seed formation and adult offspring. A high variation in the frequencies of sexual and apomictic pathways occurred at the first two stages. A trend of radical decline in realized sexuality was then observed. Sexual and apomictic seeds were produced, but the efficiency of the sexual pathway dropped drastically, and exclusively clonal offspring remained. Both reproductive pathways are unstable at the beginning of development, and only the apomictic one remains functional. Key factors reducing sexuality are the faster growth and parthenogenetic development in the aposporous pathway, and an (epi. The types of crossings that we have considered demand that we manage matings between organisms and get a large number of offspring to categorize according to phenotype in order to determine the genetic basis of a trait. This method cannot be used to analyze segregation in human individuals, and it is typically not practical for features in large domestic animals. However, occasionally it is possible to identify the method of inheritance of a feature by observing the expression of the phenotypes that represent the segregation of alleles across numerous generations of related people. This is often done using a pedigree, (Figure 2.11) which is a family tree diagram that displays the phenotype of each individual.

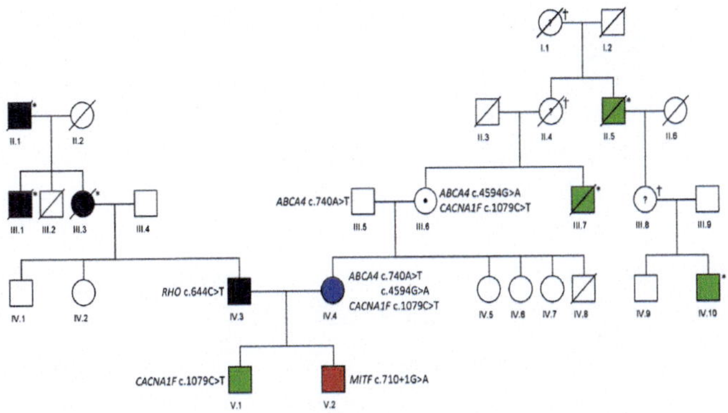

Figure 2.11. Pedigree: Genetic data of a pedigree including four different inherited retinal dystrophies. Patients with pathological findings are indicated by filled symbols. Black: *RHO*-related retinopathy. Green: *CACNA1F*-related retinopathy. Red: *MITF*-related retinopathy. Blue: *ABCA4*-related retinopathy. For individuals who underwent genetic testing, the identified variants are given next to the individual's symbol. *: affected based on family history, †: assumed carrier of *CACNA1F*variant. (Source: Birtel J, Gliem M, Hess K, Birtel TH, Holz FG, Zechner U, Bolz HJ, Herrmann P. Comprehensive Geno- and Phenotyping in a Complex Pedigree Including Four Different Inherited Retinal Dystrophies. *Genes*. 2020; 11(2):137. https://doi.org/10.3390/genes11020137, Creative Commons License).

The use of probability in pedigree analysis is a crucial aspect of genetics. When a scientist claims that a single gene with alleles P and p decides whether a pea plant's flowers will be white or purple, they do not necessarily mean that this gene is only responsible for flower color. The statement simply indicates that the recessive p mutation, which when homozygous transforms a flower's color from purple to white, has allowed scientists to identify this precise gene that determines flower color. Along with gene P, numerous other genes are required for the purple color of flowers. Among these are genes that produce the enzymes necessary for the biochemical process of creating the purple pigment anthocyanin. Rarely would a scientist with an interest in understanding the genetic basis of flower color be content with having simply discovered the P gene required for purple. The ultimate objective of a genetic analysis of flower color would be to identify at least one mutation in each gene required for purple color and, via further investigation of the mutant phenotypes, to establish the normal function of each gene that influences the trait.

2.12 MULTIPLE ALLELES

Multiple alleles are seen in two main contexts in contemporary genetics. The first is in genetic analysis, where a mutant screen may produce two or more mutant alleles for each of many genes. For instance, mutant screens were conducted in Drosophila at the beginning of the 1980s to find new recessive mutations that prevented embryonic development and thus resulted in the death of homozygous recessive embryos.

Approximately 18,000 of these mutations were discovered by the screens, and Christiane Nusslein-Volhard and Eric Wieschaus shared the 1995 Nobel Prize for their study of them with Edward B. Lewis, who had already made groundbreaking contributions to the genetics of Drosophila development (Schüpbach, 2019). Multiple alleles are also encountered by geneticists while studying natural populations of organisms. Each gene in most populations, including the human population, may have a number of alleles with marginal nucleotide sequence variations. Even if the DNA sequence of the majority of these alleles differs by one or more nucleotides, they are nonetheless able to carry out the gene's regular function and don't cause any discernible phenotypic differences. It is common for a gene to have numerous alleles in human populations. The population typically has numerous alleles of the genes utilized in DNA typing, including those used in criminal investigations. Each of these genes has a maximum of two alleles per individual, while the population as a whole frequently has 20 or more alleles. Therefore, it is unlikely that any two unrelated individuals will share the same genotype, especially if several distinct genes, each of which has several alleles, are investigated. Similar to this, examinations of affected children around the world have revealed more than 200 distinct mutated alleles of each gene for the genetic disorders cystic fibrosis and phenylketonuria. Each gene also has numerous other variants in addition to its normal version. In fact, the normal or wildtype allele for the majority of genes in the majority of populations is a collection of distinct nucleotide sequences, each of which is capable of carrying out the gene's typical activity.

In other instances, the existence of several alleles for a gene is purely accidental and reflects the history of population-level mutations that have spread among population subgroups through interbreeding and migration. In other situations, biological processes support the preservation of a large number of alleles. For instance, there can be a lot of different allelic types in the genes that regulate self-sterility in some flowering plants. Red clover species that grow wild in many fields have this kind of self-sterility. Because a pollen grain can only form a pollen tube and become fertilized if it has a self-sterility allele distinct from either of the alleles present in the flower on which it lands, the self-sterility genes inhibit self-fertilization. The pollen grain won't work on the flower if it contains an allele that is already present in the flower, in other words.

Pollen cannot operate on the same plant that produced it, and self-fertilization cannot occur because every pollen grain produced by a plant must include one of the self-sterility alleles present in the plant. Because pollen containing the new allele can fertilize all flowers except those on the same plant in these circumstances, any plant having the allele has a selective advantage. Red clover populations have hundreds of alleles of the self-sterility gene during the course of evolution, several of which have been extracted and DNA sequences identified. The fact that many of the alleles differ at several nucleotide locations suggests that the population's alleles are quite old. As demonstrated, a wildtype allele is susceptible to mutation at any of a sizable number of DNA nucleotide locations, leading to the emergence of several alleles of a gene. Some alleles in a group of many alleles could have a more profound impact on the phenotype than others. For instance, one mutant allele may totally inactivate the corresponding enzyme, whereas another mutant allele may affect the enzyme in a way that results in just a partial loss of enzyme function. Geneticists will occasionally categorize mutations based on how harmful they are.

2.13 MORE ABOUT MUTATIONS

An amorphic mutation can happen at the molecular level as a result of an amino acid substitution that renders an enzyme inactive or even a gene deletion that prevents the production of any enzyme. A hypomorph is a mutation that lowers the level of the enzyme without completely eradicating it. Usually, amino acid substitutions that reduce an enzyme's activity or stop it from being generated at a normal level cause hypomorphic mutations. A hypermorph generates more enzymes than normal, usually as a result of a mutation that alters the control of the gene in such a way that the gene product is overproduced (Magno et al., 2019). Most mutations can be categorized as amorphs, hypomorphs, or hypermorphs depending on how they change the gene's protein output. They produce none, less than, or more enzyme activity than the wildtype, nonmutant allele. But there are other kinds of mutations as well. Neomorphs are a particular kind of mutation that qualitatively change how a gene functions. For instance, a neomorph may make a gene active in a cell or tissue type where it is not typically active. Another possibility is that a neomorph will cause the expression of a gene during development at a time when the wild-type gene ordinarily isn't expressed. Neomorphic mutations in the Drosophila eyeless gene, which result in the expression of the wildtype gene product in tissues other than those that create eyes, can lead to the creation of compound eyes, replete with eye pigments, in unexpected places. Locations can include the legs, mouthparts, abdomen, or wings, basically, anywhere the wild-type eyeless gene is expressed. Ectopic expression refers to the expression of a wild-type gene at an aberrant site. An antimorph is a different kind of mutation in which the mutant gene's product competes with the healthy counterpart. In some circumstances, an amino acid substitution results in the mutant protein joining the wild-type protein to form an inactive complex. Herman J. Muller introduced these different names for mutations in 1931. Muller also observed that exposure to X-rays can result in mutations (Muller & Altenburg, 1930). Many X-ray-induced mutations involve significant disruptions or rearrangements in the DNA sequence, which cause the affected genes to express themselves in unique ways.

Genes with variable expressivity are those whose levels of expression vary across various organisms. For instance, the expression of inherited genetic diseases varies greatly from person to person in humans. One patient with the same condition might be ill, whereas another might be less badly affected. Variable expressivity describes a mutant gene that can cause one person to develop a severe version of the disease while another develops a moderate variant. Incomplete dominance refers to the fact that the phenotype predicted from a specific genotype is not always expressed, and the various degrees of expression frequently form a continuous sequence from full expression to nearly no expression of the expected phenotypic features. For instance, if a person does not smoke, even though they are genetically predisposed to lung cancer, they may not develop the illness. Environmental factors, like avoiding smoking, for instance, or the impact of other genes can cause a lack of gene expression. The expression of the trait is so weak as to be undetected in cases of incomplete dominance, which is merely the extreme of variable expressivity. The dominance of the genotype is the percentage of organisms whose phenotype matches their genotype for a particular characteristic. A genotype with 100% dominance is one that is always expressed.

2.14 GENES

Genes are microscopic elements that play a role in determining inherited features. The genes exist in pairs in higher plants and animals. Each gene pair consists of two members, one of which is inherited from the mother and the other from the father. Because DNA sequences differ, a gene might take on numerous forms. Alleles are the many varieties of a gene. The genotype of an organism is the unique set of gene alleles that it possesses. An organism's phenotype is made up of its visible traits. In an organism, the genotype is homozygous for the A or an allele if the two alleles of a gene pair are the same (for instance, AA or aa); otherwise, the genotype is heterozygous (Aa). The allele that is expressed is referred to as dominant, and the allele that is hidden is referred to as recessive, when the phenotype of a heterozygote is the same as homozygous genotypes. The F1 generation, which is used in genetic studies, consists of the offspring of a cross. The F2 generation is created when F1 generation members mate. The ratio of genotypes in the F2 generation in a cross like AA x aa, in which only one gene is taken into account (a monohybrid cross), is 1 dominant homozygote (AA): 2 heterozygotes (Aa): 1 recessive homozygote (aa). The members of each allelic pair separate into various gametes during gamete production, which leads to the Mendelian ratios of genotypes and phenotypes, and the random union of gametes during fertilization. Probability laws govern the segregation, independent selection, and random union of gametes, which form the foundation for forecasting the results of genetic crosses. The addition rule and the multiplication rule are the two fundamental formulas for combining probability. According to the additional rule, which only applies to events that cannot coexist, the likelihood that 2 events will occur is equal to the product of their individual probabilities. According to the multiplication rule, which applies to independent events, the likelihood that two independent events will occur simultaneously is equal to the sum of their individual probabilities. Controlled crosses cannot be performed on some organisms, such as humans, thus genetic analysis is instead carried out by looking at several generations of a family tree known as a pedigree. A pedigree's potential genotypes and the probability that each family member has a specific genotype are determined through pedigree analysis.

The functional definition of a gene is the complementation test (Stahl & Murray, 1966). If a homozygous recessive mutation crosses with another recessive mutation to produce nonmutant offspring, then the two mutations are regarded as alleles of distinct genes. These alleles are said to work well together. On the other hand, if a homozygous recessive mutation crosses with another recessive mutation to produce mutant offspring, the two recessive mutations are regarded as alleles of the same gene. These alleles are said to be complement-failures. Crossing the homozygous recessives in all pairwise pairings is required for a comprehensive complementation test for each collection of recessive mutations. As a result of mutant screens or in natural populations, several alleles are frequently found. There are more than two different variants of a gene when there are multiple alleles. The genes employed in DNA typing and the alleles that cause self-sterility in some flowering plants are two examples of alleles with a high number. Despite the possibility of numerous alleles in a population, only one allele of each gene can be carried by a gamete and a maximum of two different alleles of each gene can be carried by an organism. Carl Nageli's discovery that the nucleus divides first in dividing cells in the 1840s provided the first evidence of the significance of the cell nucleus and its components. Nageli likewise missed the significance of nuclear division. He considered the cells where he observed nuclear division to be abnormal. But by the 1870s, it was understood that nuclear division is a characteristic of cell division that is present in all cells. The revelation that the nuclei of two gametes fuse during fertilization occurred almost simultaneously with the discovery of the relevance of the nucleus in inheritance. The finding of chromosomes, which had been made visible by light microscopy when stained with simple dyes, was the next significant development ten years later. A few years later, it was discovered that chromosomes partition in an ordered manner into the gametes created by the division of reproductive cells as well as the daughter cells created by cell division.

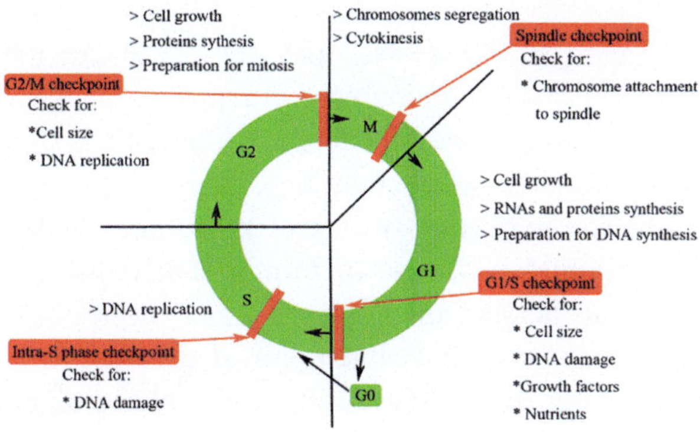

Figure 2.12. Cell cycle. (Source: Wu Y-S, Liang S, Li D-Y, Wen J-H, Tang J-X and Liu H-F (2021) Cell Cycle Dysregulation and Renal Fibrosis. Front. Cell Dev. Biol. 9:714320. doi: 10.3389/fcell.2021.714320, Creative Commons License).

2.15 CELL CYCLE

Genetics regulates the cell cycle itself. All eukaryotes appear to have roughly the same regulatory mechanisms. The transitions from G1 into S and from G2 into M are also crucial. The G1/S and G2/M transitions are referred to as checkpoints because they are postponed until certain crucial processes are finished. For instance, for DNA replication to begin at the G1/S checkpoint, either enough time must have passed since the previous mitosis (in some cell types) or the cell must have grown sufficiently (in other cell types).

Similar to the G2/M checkpoint, the M phase cannot begin unless DNA replication and any DNA damage repair are finished (Figure 2.12). Both key control points are controlled similarly and employ a specific protein kinase that controls the activity of target proteins via phosphorylation, known as the p34 kinase subunit. One of the many varieties of protein kinases that are employed to control biological activities is the p34 kinase. The p34 polypeptide component must combine with a number of other polypeptide chains known as cyclins because their abundance cycles in time with the cell cycle in order to become active. One pair of cyclins joins with the p34 subunit at the G1/S checkpoint to produce the active kinase that initiates DNA replication and other S phase processes. Similar to this, at the G2/M checkpoint, a second group of cyclins join with the p34 subunit to produce the active kinase that starts the chromosomal condensation, opening of the nuclear membrane, and rearrangement of the cytoskeleton in preparation for cytokinesis.

Prophase, Metaphase, Anaphase, and Telophase are the 4 stages of mitosis that are traditionally recognised. In prophase, the chromosomes in interphase take on the shape of stretched filaments and are not visible under a light microscope as distinct entities. The nucleus appears scattered and granular, with the exception of one or more noticeable black masses (nucleoli). Chromosomes condense to form visually distinct, thin threads within the nucleus at the start of prophase. Later in prophase, it is simple to discern that each chromosome. Each pair of chromatids is the result of one chromosome being duplicated during the S phase of interphase. The centromere is a specific area of the chromosome where the chromatids in a pair are held together. The intricate coiling of the chromosomes causes them to become shorter and thicker as prophase develops. The nucleoli vanish and the nuclear envelope, a membrane enclosing the nucleus, quickly degrades after the conclusion of prophase.

In metaphase, the mitotic spindle develops at the start of the metaphase. The spindle is a bipolar structure made up of microtubule bundles that resemble fibers that extend across the cell in between the spindle's poles (Loughlin et al., 2011)but mechanisms determining spindle length are poorly understood. Based on a 2D model of meiotic spindle assembly, we predicted that higher localized microtubule (MT. In the region of the centromere, each chromosome joins with a number of spindle fibers. The kinetochore is the name given to the centromere-associated structure to which the spindle fibers are connected. The chromosomes travel toward the cell center after being joined to spindle fibers until all of the kinetochores are evenly spaced from the spindle poles. The metaphase plate is the name of this hypothetical plane. The chromosomes attain their maximum contraction when they are aligned on the metaphase plate, making it simple to count them and look for morphological variations. In both mitosis and meiosis, proper chromosomal alignment is a critical cell cycle regulatory checkpoint at metaphase. The end of metaphase is postponed in a cell where a chromosome is only connected to one pole of the spindle. The kinetochore is the source of the signal for chromosome alignment, and it appears that the chemical basis of the signal is the dephosphorylation of specific kinetochore-associated proteins. The discovery that an unattached chromosome whose kinetochore has been damaged by a concentrated laser beam does not cause metaphase to be delayed serves as evidence for the kinetochore's function. The use of an antibody that only reacts with certain kinetochore proteins when they are phosphorylated allows for the demonstration of the role of dephosphorylation.

Kinetochores that are not connected to the spindle diminish the response but combine strongly with the antibody. The antibody response with the kinetochore recurs in chromosomes that have been severed from the spindle. The metaphase checkpoint is passed and the cell continues to divide when all of the kinetochores are under tension and aligned on the metaphase plate through the signaling mechanism.

In anaphase, each chromosome's two sister chromatids travel to the spindle's opposing poles as the centromeres split longitudinally. Each sister chromatid that has undergone centromere division is treated as a distinct chromosome. The spindle fibers connected to the centromeres gradually become shorter, which causes the chromosomes to migrate in different directions toward the poles. The chromosomes are located in two groups close to the spindle's opposite poles at the end of anaphase. The same number of chromosomes that were in the initial interphase nucleus are present in each group. In telophase, each compact set of chromosomes produces a nuclear envelope around them during telophase, nucleoli are created, and the spindle is lost. The condensation of the chromosomes is reversed until they are no longer discernible as separate objects. Meiosis is a type of cell division where only one copy of each pair of chromosomes found in the premeiotic cell is found in the newly formed cells. During meiosis, a diploid cell with two sets of chromosomes divides into four daughter cells, each of which is genetically distinct and has one haploid set of chromosomes. Two successive nuclear divisions make up meiosis. Each pair of chromosomes becomes closely connected along its length before the first nuclear division. Homologous chromosomes are those chromosomes that pair with one another. Each member of a homologous chromosome is made up of two sister chromatids connected at the centromere because each member has already undergone replication. The homologous chromosome pairing results in a four-stranded structure as a result. During the initial nuclear division, homologous chromosome pairs are split apart, with one member of each pair moving to the spindle's opposing poles. There are two nuclei created, each with a pair of duplex chromosomes that are haploid.

There is no chromosomal replication during the second nuclear division, despite the division roughly resembling a mitotic division (McGrew et al., 1992). The chromosomes get arranged on the metaphase plate during metaphase, and during anaphase, each chromatid splits into two opposing daughter nuclei. Four haploid daughter nuclei, each having the equivalent of one sister chromatid from each pair of homologous chromosomes, are produced as a result of meiosis' two divisions. Through the exchanges, regions from one homologous chromosome are mixed with portions from the other to create chromosomes. Meiosis occurs in specific cells in animals known as meiocytes, which is a collective word for the primordial oocytes and spermatocytes in the gamete-forming organs. Both spermatocytes and oocytes develop into egg and sperm cells, respectively. In plants, Meiotic products often take the shape of spores, which go through one or more mitosis cycles to create a haploid gametophyte organism. The haploid nucleus of the gametophyte undergoes mitotic division to produce gametes. When haploid gametes fuse, a diploid zygote is produced, which grows into a sporophyte plant and goes through meiosis to produce spores, restarting the cycle. Meiosis typically takes days or even weeks and is far more difficult and time-consuming than mitosis. The main point is that meiosis involves two nuclear divisions but only one chromosome duplication. The first and second meiotic divisions of the nucleus can be divided into a series of stages that are comparable to those used to define mitosis.

Chapter

3 INTRODUCTION TO GENETIC ENGINEERING

CONTENTS

3.1 INTRODUCTION

Any scientific discipline's potential to advance depends on the availability of methodologies and techniques that increase the variety and level of experimentation possible. The advent of genetic engineering in the last 35 years or so has spectacularly shown this. This area has developed quickly to the point where it is now standard procedure in many laboratories across the world to isolate a specific DNA fragment from an organism's genome, determine its base sequence, and evaluate its function. The technology is currently utilized in a wide range of other fields, such as forensic analysis of crime scene samples, paternity cases, medical diagnostics, genome mapping and sequencing, and the biotechnology sector. The technology of gene manipulation is particularly remarkable because it is easily available to individual scientists without the requirement for expensive machinery or resources that are outside the purview of a research institution with enough funding (Musiol-Kroll et al., 2019)clinically used antibiotics and other valuable bioactive products, as well as biotechnologically relevant enzymes. Most strains were discovered by their ability to produce a given molecule and were often poorly characterized, physiologically and genetically. The development of genetic methods for Streptomyces and related filamentous actinomycetes has led to the successful manipulation of antibiotic biosynthesis to attain structural modification of microbial metabolites that would have been inaccessible by chemical means and improved production yields. Moreover, genome mining reveals that actinomycete genomes contain multiple biosynthetic gene clusters (BGCs. The bioscience community as a whole still has access to the technology in some way, even though it has recently grown significantly as a result of the establishment of genome sequencing programs. Single genome-wide DNA sequencing can be thought of as a set of four processes that make up the essence of gene cloning. The genetic engineer receives a specific DNA sequence after successfully completing these processes, which can then be applied to various projects. Gene cloning can be seen as a type of molecular agriculture, allowing for the manufacture of vast quantities of a specific DNA sequence. This ability to isolate a specific gene sequence remains a key component of gene manipulation used on a daily basis in research labs around the world, even in the era of large-scale sequencing initiatives.

Along with technological and scientific difficulties, one feature of the new genetics that has raised concerns is that it raises many moral and ethical issues. Genetics is a concept that has been developed to characterize the ethical issues that contemporary genetics faces, which are projected to grow in quantity and complexity as genetic engineering advances. Gene therapy, the use of transgenic plants and animals, the study of the human genome, and many other topics are of relevance to the general public as well as scientists. A popular reaction against the technology has been sparked by recent breakthroughs in genetically modified foods. Advancements in cloning living organisms, as well as in fields like *in vitro* fertilization and xenotransplantation, pose more concerns. Another area of concern related to the general advancement in genetic technology is stem cell research and the possible medicinal benefits that this research may offer. It remains to be seen if we can apply genetic engineering for the overall benefit of humanity and prevent the technological misuse that frequently follows scientific progress, taking into account all the potential costs and advantages. Although gene manipulation techniques are relatively new, it is important to keep in mind that knowledge and experience of microbiologists were crucial in their development. The rediscovery of Gregor Mendel's work at the turn of the century marked the true beginning of genetics, and over the course of the following 40 years or more, the concepts of heredity and genetic mapping were clarified. In the middle of the 1940s, the field of microbial genetics was founded, and DNA's function as genetic material was confirmed. The methods of gene transfer across bacteria underwent significant advancements during this time, and substantial information was built upon which subsequent improvements might build.

3.2 GENETICS AS THE BASE

The molecular level growth of genetics was improved by James Watson and Francis Crick's discovery of the DNA structure in 1953, and the following few years witnessed a period when DNA was possibly the most renowned and most readily recognizable molecule in the entire world. The main characteristics of the gene and its expression were the subject of activity and enthusiasm. The culmination of this effort was the development of the full genetic code in 1966; this prepared the way for the advent of the new fields of genetics. Research has advanced to the point where technological challenges are impeding advancement. It was not possible to extend the genetic coding to further study the gene. However, a series of innovations gave gene modification the impetus it needed to become a reality. The enzyme DNA ligase was discovered in 1967 (Tomkinson et al., 2006). This enzyme can act as a kind of molecular glue by joining two strands of DNA together, which is necessary for the creation of recombinant molecules. The first restriction enzyme (Figure 3.1) was then isolated in 1970, marking an important turning point in the history of genetic engineering. In essence, restriction enzymes are molecular scissors that cut DNA at specific predetermined sequences. These enzymes can be used to create DNA fragments that can be joined to other DNA fragments. As a result, the fundamental technologies needed to create recombinant DNA were available by 1970. In 1972, researchers at Stanford University used the ability of DNA ligase to link DNA strands together and the cleavage capabilities of restriction enzymes (scissors) to create the first recombinant DNA molecules. It is impossible to overstate the significance of these initial, unsuccessful experiments. Now that distinct DNA molecules could be joined together, scientists might connect the DNA of one organism to that of an entirely different organism. Scientists used the plasmid pSC101, which was obtained from the bacterium *Escherichia coli* in 1973 by connecting DNA fragments to it, to make sure that the target sequence is replicated in an appropriate host cell. These recombinant molecules have replicon behavior, which allows them to multiply when added to *E. coli* cells. Thus, specific DNA fragments from bacterial colonies that formed clones when grown on agar plates could be isolated by creating recombinant molecules *in vitro* and putting the construct in a bacterial cell where it could replicate *in vivo*. This advancement signaled the beginning of the field of science known as gene cloning.

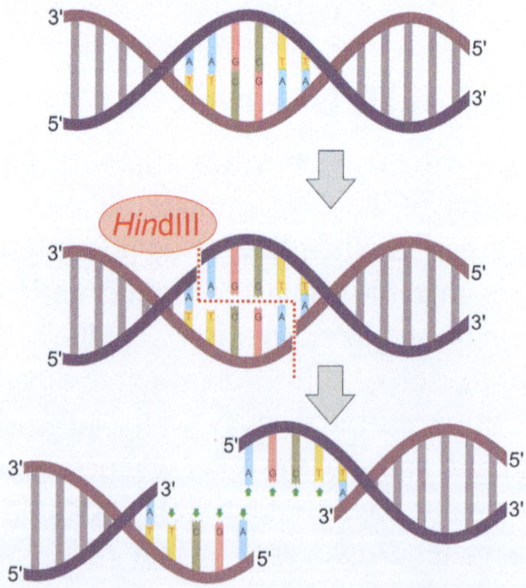

Figure 3.1. Restriction enzyme. (Source: By Helixitta - Own work, CC BY-SA 4.0, https://commons.wikimedia.org/w/index.php?curid=41966112, Creative Commons License).

Modern genetics, which is arguably the biggest scientific revolution of all, was brought about by the findings of 1972 and 1973. The new technology was widely used, and there was a strong sense of urgency and enthusiasm. The issue about the safety of these organisms and the effects of releasing GMOs into the environment has been reopened by the introduction of genetically modified organisms (GMOs), particularly crop plants. It can be helpful to consider what cells are and how living systems are organized before we look at the molecular biology of the cell. In biological systems, structure and function are closely related to one another. The structure and function of living systems are highly interdependent and are organized hierarchically. Components are combined in more intricate patterns, resulting in the emergence of new functional features. Biological system chemistry is based on the element carbon. By combining smaller molecules using the dehydration synthesis method, which can establish four covalent connections with additional atoms, complex molecules (macromolecules) are created. Building forming molecules, which may then be combined together to form macromolecules requires linking carbon atoms together and adding other atoms. Lipids, carbohydrates, nucleic acids, and proteins, are the four categories of macromolecules that biologists typically recognize (Abarca-Cabrera et al., 2021). A condensation reaction between functional groups on the molecules to be brought together occurs during the production of macromolecules. Through the removal of the molecules of water, this dehydration synthesis creates a covalent bond. Many monomeric units may be brought together in this manner to form the polymeric macromolecules of the cell, such as polysaccharides, proteins, and nucleic acids. By reintroducing water, the polymers can be disassembled into their constituent monomers. This is referred to as hydrolysis.

3.3 BIOLOGICAL SYSTEMS

The fundamental unit of organization in biological systems is the cell. There are several characteristics that all cells share, despite the fact that there are many different types of cells. The interface between the interior of the cell and the surrounding environment is a cell membrane called the plasma membrane. A cell wall may also be present in some cells, including those of bacteria, yeasts, and plants, adding further structural support. The basic unit of biological systems is the cell, which must contain some form of genetic material and give the information necessary for cells to operate. Eukaryotic cells have nuclei, whereas prokaryotic cells have not. One method of cell classification is provided by the organization of this genetic information. DNA is not compartmentalized in prokaryotic cells (bacteria), but it is housed inside a membrane-bound nucleus in eukaryotic cells. Membranes are also used by eukaryotic cells. All cells have a maximum upper size limit, albeit prokaryotic cells are typically smaller than eukaryotic cells. In multicellular eukaryotes, the size of the organism is increased by producing more cells rather than by enlarging the cells (Kirk, 2005)a genus of spherical organisms that have thousands of cells and a germ–soma division of labor. It is estimated that these organisms all shared a common ancestor about 50 ± 20 MYA. Here we outline twelve important ways in which the developmental repertoire of an ancestral unicell similar to modern C. reinhardtii was modified to produce first a small colonial organism like Gonium that was capable of swimming directionally, then a sequence of larger organisms (such as Pandorina, Eudorina and Pleodorina. It helps to imagine that the four nitrogenous bases adenine (A), guanine (G), cytosine (C), and thymine (T) are responsible for directing life.

Proteins, especially the enzymes that catalyze metabolic events, are ultimately responsible for the expression of genetic information. Given that a protein can have several hundred amino acid residues, the variety of proteins that can be produced is practically limitless, presuming that the right amino acid sequence can be determined from genetic information. We may estimate that using the bases alone would not give enough scope to encode 20 amino acids because there are only 4 viable code combinations because the bases are essential basal components (A, G, C, and T). It would still not be enough if the bases were organized in pairs, which would result in 42 or 16 potential combinations. 43 or 64 different permutations are offered by triplet combinations, which is more than enough (Anderson & Schultz, 2003)it has been shown that an amber suppressor tRNA/aminoacyl-tRNA synthetase pair derived from thc tyrosyl-tRNA synthetase of Methanococcus jannaschii can be used to genetically encode unnatural amino acids in response to the amber nonsense codon, TAG. However, we have been unable to modify this pair to decode either the opal nonsense codon, TGA, or the four-base codon, AGGA, limiting us to a 21 amino acid code. To overcome this limitation, we have adapted a leucyl-tRNA synthetase from Methanobacterium thermoautotrophicum and leucyl tRNA derived from Halobacterium sp. NRC-1 as an orthogonal tRNA-synthetase pair in Escherichia coli to decode amber (TAG. Thus, by designating the amino acids using sets of three nucleotides (codons), a simple coding scheme may produce a wide variety of protein forms and functions. As a result, a strand of DNA would need to include 900 nucleotides in order to code for a protein with 300 amino acids. Despite the fact that there are 64 potential codons rather than only 20, three of them are STOP codons. Wobble is a different word for the same situation, in which the first two bases of a codon are frequently crucial but the third is not. Messenger RNA (mRNA) serves as an intermediary in the unidirectional transport of genetic information from DNA to proteins. Transcription (TC) is the process of genetic information from DNA being copied into RNA. The processes of transcription and translation are involved in the movement of genetic information from DNA to RNA to proteins.

3.4 GENE EXPRESSION

All investigations of gene expression share this idea of information flow, also referred to as the Central Dogma of molecular biology.

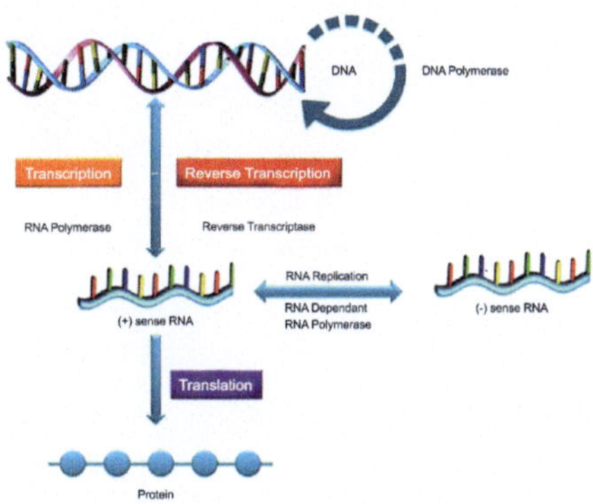

Figure 3.2. Reverse transcriptase. (Source: Saxena, S. K., & Chitti, S. V. (2016). Molecular Biology and Pathogenesis of Retroviruses. InTech. doi: 10.5772/62885, Creative Commons License).

Reverse transcriptase, (Figure 3.2) an RNA-dependent DNA polymerase, is an enzyme that certain viruses possess. It converts the single-stranded RNA genome into a double-stranded DNA molecule. Double-stranded DNA makes up the majority of the genetic material in most organisms. In order for genetic information to work in a live organism for up to 100 years or more, it must first be stable (Sivanandam & Deepa, 2008). To enable the spread of genetic information when new cells are generated during growth and development, the molecule must also be able to replicate. In order for evolutionary pressures to have an impact, there should be a chance for small genetic changes (mutation). A chain of nucleic acids is frequently referred to as a polynucleotide since they are heteropolymers made of nucleotide-based monomers. Three elements, sugar, phosphate group, and nitrogenous base make up each monomer individually. The two nucleic acids, deoxyribonucleic acid (DNA) and ribonucleic acid (RNA) are polymers made of nucleotides.

The nitrogenous bases dictate the structural properties of polynucleotides, and the sugar/phosphate components of a nucleotide determine how well they store and transmit information. A polynucleotide can acquire directionality when nucleotides are linked together by a 5'-3' phosphodiester linkage. As a result, the molecule will contain a free -OH or hydroxyl group at the 3' end and a free phosphate group at the 5' end, which has significant ramifications for the structure, operation, and manipulation of nucleic acids. The sugar-phosphate chains are found in an antiparallel arrangement in a double-stranded molecule like DNA, with the two strands running in opposing directions.

Regarding their role in coding, nitrogenous bases are crucial parts of nucleic acids. The complementary base pairing of the nucleotides A, T, and G in DNA is essential for the storage, transmission, and utilization of information. Thymine is substituted with the functionally identical nucleotide uracil (U) in RNA. Chemically, cytosine, thymine, and uracil are pyrimidines, which have a single ring structure, whereas adenine and guanine are purines, which have a double-ring structure.

The only acceptable pairing is a purine—pyrimidine base pair, which is determined by the bonding configurations of the atoms in the bases as well as the space limitations of the DNA molecule. Hydrogen bonds, two in the case of an A-T base pair and three in the case of a G-C base pair hold the bases together.

The DNA molecule often takes the B-form, a right-handed double helix. Watson and Crick suggested this structure in 1953. The Z-form and the A-form (right-handed helix) are alternate forms of DNA.

3.5 NUCLEIC ACIDS

The key chemical distinctions between RNA and DNA are the existence of ribose rather than 2-deoxyribose and uracil rather than thymine. Although short segments of double-stranded RNA can occasionally be discovered in self-complementary areas, RNA is mostly single-stranded. Ribosomal RNA (rRNA), messenger RNA (mRNA), and transfer RNA (tRNA) are three major forms of RNA that are present in cells (Okamura et al., 2015)

RNAs are transcribed in the nucleus and exported to the cytoplasm through the nuclear pore complex. The RNA molecules that are exported from the nucleus into the cytoplasm include messenger RNAs (mRNAs.

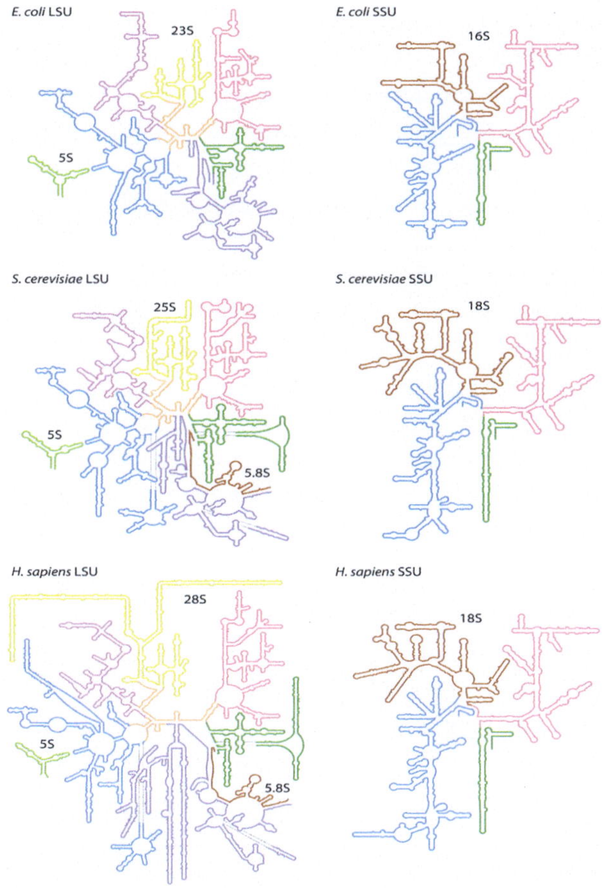

Figure 3.3. ribosomal RNA molecules of many species. (Source: https://en.wikipedia.org/wiki/File:T_thermophilus_S_cerevisiae_H_sapiens.png, Creative Commons License).

The most prevalent type of RNA molecule is ribosomal RNA, (Figure 3.3) which accounts for about 85% of all cellular RNA. It is connected to ribosomes, a crucial component of the translational apparatus. About 10% of all RNA is made up of transfer RNAs, which offer the crucial specificity required for the insertion of the right amino acid into the protein that is being synthesized. Less than 5% of all cellular RNA is typically messenger RNA, which serves as the genetic information carrier from the DNA to the translational machinery. The fundamental building block of genetic information is the gene. The study of genes was mostly indirect until molecular biology and the discovery that genes were made of DNA; the effects of genes were seen in phenotypes, and the behavior of genes was studied. Despite the apparent drawbacks of this strategy, a wealth of knowledge regarding the operation of genes was collected, and the fundamental principles of transmission genetics were developed. The nomenclature used in this field of genetics expanded as the gene was researched in greater depth, and theories about genes were revised to account for new information. The term gene is typically understood to refer to the genetic data that has been translated into a single RNA molecule. The gene is the fundamental building block of a single protein. Genetic information for RNA molecules is an exception. A specific genetic region on chromosomes is where genes are found. Alleles are distinct variations of a gene, such as rRNA and tRNA, which are not translated. Prokaryotic cells also have a slightly distinct nomenclature because of the way their genes are set up. Chromosomes are where genes are found, and the area of the chromosome where a certain gene is situated is known as the locus of that gene. Different variations of the same gene are referred to as alleles in diploid organisms, which have their chromosomes organized as homologous pairs.

Although there is no such thing as a typical gene, every gene must meet a few fundamental conditions in order to work. Although most species only use one strand to encode each gene, the double-stranded DNA molecule has the capacity to store genetic information in either strand. The terms used to describe the two DNA strands, such as coding/non-coding, sense/antisense, plus/minus, transcribed/non-transcribed, and template/non-template, have the potential to cause confusion. It might be confusing when various authors use the same phrases in different contexts. The names coding and non-coding are preferred by the International Union of Biochemistry and the International Union of Pure and Applied Chemistry, with the coding strand of DNA being understood to be the mRNA-like strand. When DNA strands are copied during cloning methods, such as when no coding function is necessarily involved, the phrases template and non-template will be used to characterize the DNA strands. Thus, the non-coding strand of DNA is translated into an mRNA molecule, which has the same sequence as the coding strand of DNA, to express genetic information. When dealing with DNA sequence information, the coding strand's sequence is typically presented because it makes it simple to compare it to the RNA sequence. There are additional regulatory sequences linked to genes in addition to the base sequence that determines the codons in a protein-coding gene. It is necessary to have a location where transcription can begin. This location includes the promoter (P), which binds RNA polymerase, as well as a specific transcriptional initiation site or TC. Additionally needed is a stop site for transcription or tC. The DNA region that is transcribed into RNA is known as the transcriptional unit and runs from TC start to tC stop (Grummt, Kuhn, Bartsch, & Rosenbauer, 1986)one of which has been identified approximately 2 kb upstream of the pre-rRNA start site. The interaction of a specific nuclear factor with the upstream terminator increases the efficiency of initiation, suggesting that transcription termination and initiation at the adjacent promoter work in an interrelated fashion.","container-title":"Cell ","DOI":"10.1016/0092-8674(86. Genes contain a number of crucial areas within this transcriptional unit. A promoter is required for RNA polymerase binding, and the transcriptional unit is defined by the start and stop sites. It is possible for additional sequences to exist either upstream or downstream from the gene itself and have a role in the regulation of gene expression.

3.6 GENES

Genes are typically found arranged in groups called operons in prokaryotic cells like bacteria. The operon is a group of linked genes. Prokaryotic genes are frequently found in operons, which are collections of related genes that are each controlled by a single promoter/regulatory region.

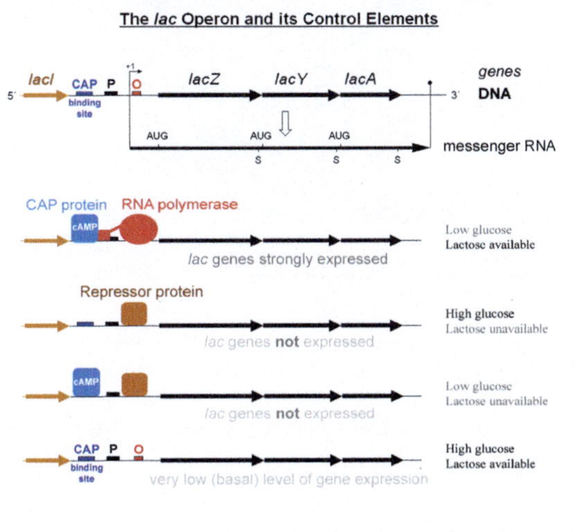

Figure 3.4. The lac operon and its working. (Source: https://commons.wikimedia.org/wiki/File:Lac_operon-2010-21-01.png, Creative Commons License).

These operons frequently contain genes that code for enzymes involved in metabolic pathways. The lac operon, which codes for the enzymes involved in lactose catabolism, is possibly the best-known illustration of this operon.

Three structural genes, i.e., genes that encode proteins are contained within the operon, (Figure 3.4) along with the operator, an upstream control region that includes the promoter and a regulatory site. The repressor gene, which is located outside the operon, produces the Lac repressor protein, which binds to the operator site and controls the operon negatively by preventing RNA polymerase from binding. The translated mRNA may contain information for more than one protein due to the frequent grouping or clustering of structural genes in prokaryotes. A polycistronic mRNA is one of these molecules, with cistron standing for the gene i.e., encoding one protein. Therefore, a large portion of a bacterium's genetic material is expressed through polycistronic mRNAs, whose synthesis is controlled by the requirements of the cell at any particular time. This system, which is adaptable and effective, enables the cell to quickly adjust to shifting environmental conditions.

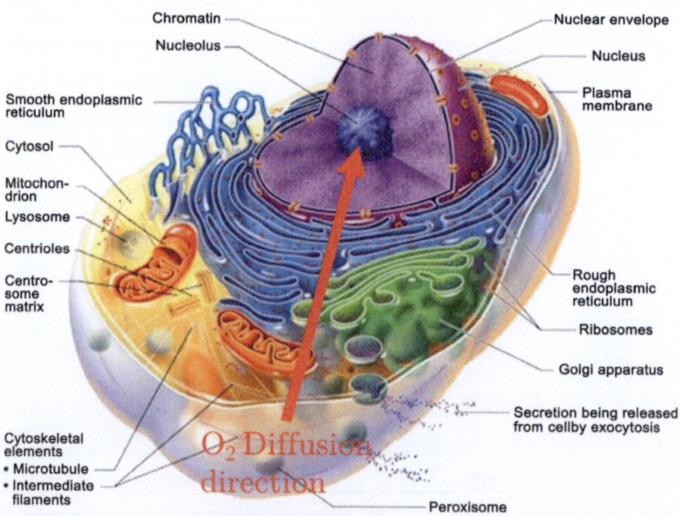

Figure 3.5. The importance of the nucleus. (Source: Seco J, King CC, Camazzola G, Jansen J, Tirinato L, Marafioti MG, Hanley R, Pagliari F, Beckman SP. Modulating Nucleus Oxygen Concentration by Altering Intramembrane Cholesterol Levels: Creating Hypoxic Nucleus in Oxic Conditions. *International Journal of Molecular Sciences*. 2022; 23(9):5077. https://doi.org/10.3390/ijms23095077, Creative Commons License).

The existence of a membrane-bound nucleus (Figure 3.5), where the DNA is housed in the form of chromosomes, is a key characteristic that distinguishes eukaryotic cells from other types of cells. Thus, translation takes place in the cytoplasm, while transcription takes place inside the nucleus. The inclusion of genetic information in mitochondria and chloroplasts, which have their own distinct genomes that specify many of the components required by these organelles, further complicates the situation. Because of the significant effects this compartmentalization has on genetic and metabolic control, eukaryotic gene structure and function are more complex than those of prokaryotes.

3.7 MRNA

In 1977, it was discovered that eukaryotic genes contained extra DNA fragments that did not show up in the mRNA that the gene encoded (Carlson & Botstein, 1982)glycosylated form, the synthesis of which is regulated by glucose repression, and an intracellular, nonglycosylated enzyme that is produced constitutively. The SUC2 gene has been cloned and shown to encode two RNAs (1.8 and 1.9 kb.

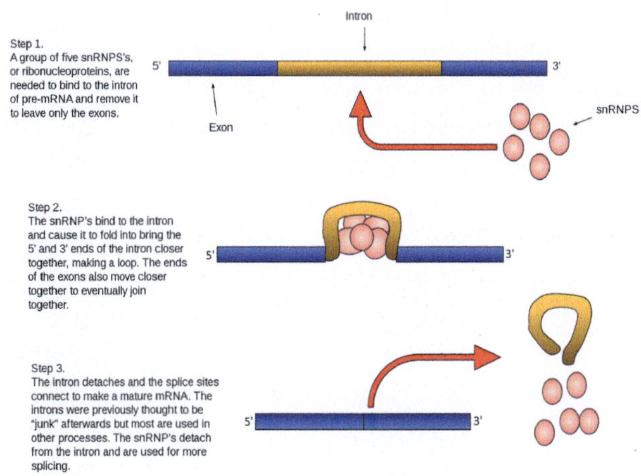

Figure 3.6. Introns and the process of splicing. (Source: https://upload.wikimedia.org/wikipedia/commons/3/37/Process_of_RNA_splicing.png, Creative Commons License).

The sequences that will make up the mRNA are referred to as exons, and these sequences are known as intervening sequences or introns. As in the chicken ovalbumin gene, which is typical of eukaryotic genes and has a total of seven introns making up more than 75% of the gene, the number and length of the introns frequently outweigh that of the exons. The main transcript, which is transformed by RNA processing into the mature mRNA, contains the introns (Figure 3.6).

As our understanding has grown, it has become evident that eukaryotic genes are frequently exceedingly complicated and may even be rather massive. This demonstrates the large range of sizes for human genes, with the lowest perhaps only being a few hundred base pairs long. On the X chromosome, the dystrophin gene is distributed over 2.4 Mb of DNA, with the 79 exons accounting for just 0.6 percent of this length of DNA. Since the introns must be deleted before the mRNA can be translated, the existence of introns clearly has significant effects on how genetic information is expressed in eukaryotes. The introns of the main transcript are spliced out in the nucleus during this process. The inclusion of a cap at the 5' end and an adenine tail at the 3' end are two intranuclear modifications. A fully functional mRNA that is prepared for export to the cytoplasm for translation is the result of these alterations, which are a component of the process known as RNA processing.

Perhaps there is a purpose to genetic information. The term genetic blueprint is occasionally used, however, It is preferable to think of it as a recipe than as a blueprint. Depending on the type of mutation and where it occurs in the gene sequence, the effects of a mutation can be moderate or severe. Mutations in non-essential regions of genes may not have any effects, whereas mutations in key parts of genes may have very negative effects. A single base insertion or substitution can occasionally have a significant impact. The information to specify the proteins is provided by genes, but for the cell to function properly, the entire process must be controlled and regulated. The crucial phases in creating functioning proteins in the cell are these two processes. The process of transcription entails creating an RNA from the DNA template that the concerned transcriptional unIt is non-coding strand has provided. There is only one RNA polymerase enzyme in prokaryotes, while there are three different varieties in eukaryotes (I, II, and III). These, in turn, synthesize messenger, transfer, and ribosomal RNAs. Large multi-subunit proteins with relative molecular weights of about 500000 make up all RNA polymerases (Travers, 1993). The phases of transcription include binding of the DNA/RNA polymerase, chain start, chain elongation, chain termination, and release of the RNA. When the RNA molecule is released, it may be processed and exported to the cytoplasm (as in eukaryotes) before translation takes place, as in prokaryotes, or it may be instantly available for translation (as in prokaryotes). An mRNA molecule, a supply of charged tRNAs (tRNA molecules with their associated amino acid residues), and ribosomes are necessary for translation. Protein synthesis takes place in the ribosomes in prokaryotes, where the codon/anticodon recognition event establishes the connection between nucleic acids and proteins.

3.8 RRNA

Three rRNAs and about 52 distinct ribosomal proteins make up ribosomes. The ribosome is a complicated structure that essentially functions to hold the mRNA in place while the codons are matched with the proper anticodon on the tRNA, ensuring that the right amino acid is added to the expanding polypeptide chain. The polypeptide elongation from the N terminus to the C terminus is represented by the translation of the mRNA molecule in a 5'-3' orientation (Liu et al., 2002)elongation factor 1α (EF1α. The processes through which genes are expressed are provided by transcription and translation. However, it is crucial to regulate gene expression so that the proper gene products are created in the cell at the right time. Gene regulation in prokaryotes frequently happens in response to environmental cues like nutrition availability. Bacterial cells must be able to adapt to a wide range of environmental situations, thus they must always have all of their genetic material available in case certain gene products are required. Bacteria that maintain this state of readiness in their genomes can conserve energy and react promptly to any changes in the food supply. This is an illustration of adaptive gene expression regulation. Human cells encounter a substantially distinct set of environmental variables than microorganisms. Eukaryotic genes are frequently controlled in response to signals produced by the organism itself. Cell specialization hence leads to more complicated functions but necessitates more controlled environments. Since development is a result of differentiation, multicellular eukaryotes frequently have controlled genes. As one might anticipate, intricate gene regulation occurs throughout an organism's growth and life cycle.

Many instances of gene products, in addition to those that are regulated and controlled, are present throughout a cell's lifespan. Since these genes are essentially uncontrolled and encode proteins that are always necessary, they are also known as housekeeping genes or constitutive genes (Eisenberg & Levanon, 2003)the set is expected to be rich in 'housekeeping' genes, showing constitutive expression in all tissues. We compare selected aspects of their genomic structure with a set of background genes. We find that the introns, untranslated regions and coding sequences of the housekeeping genes are shorter, indicating a selection for compactness in these genes.","container-title":"Trends in Genetics","DOI":"10.1016/S0168-9525(03. A bacterial cell will encounter a variety of environmental situations while it is living outside of the lab. There will be variations in the availability of nutrients, in particular. In order for the cell to survive, energy resources must be conserved, hence unnecessary protein production must be avoided. As a result, bacterial cells contain systems that allow for highly sensitive control of operons. An operon that encodes proteins engaged in a catabolic pathway, which breaks down materials to release energy, is frequently controlled by turning on only when the substance is beginning to be produced. In contrast to an on/off switch, gene regulation in bacteria allows a range of levels of gene expression to be achieved. As a result, systems exist that keep catabolic operons off when the material is absent. These are referred to as inducible operons and are often under the control of a negative control mechanism including a repressor protein that blocks RNA polymerase from accessing the promoter. The lac operon is a main example of a catabolic operon. The repressor protein attaches to the operator in the absence of lactose, turning the system off. However, because of a minor leak, the proteins expressed by the operon will only be minimally present in the cell. When lactose is made available, the permease transports it inside the cell where it attaches to the repressor protein, changing its conformation and preventing it from binding to the operator. As a result, the operon is now accessible to RNA polymerase and the negative control has been eliminated. Only when lactose is available and energy levels are low can complete activity occur, due to a second level of control based on the concentration of cAMP. This dual-control mechanism is a particularly efficient method of controlling gene expression since it allows for a variety of expression levels that are more akin to a weak switch than an on/off switch. This guarantees that the enzymes are only synthesized at their highest rate when they are actually needed in the case of catabolic operons like the lac system.

3.9 GENE STRUCTURE

Gene structure was obviously one of the first areas where efforts were focused when methods for examining DNA were developed (Kuo & Allis, 1999). The entire genome of an organism is known as the genome. The focus of molecular biology has gradually evolved in recent years, and today we are far more likely to think of the genome as a whole, almost as a form of cellular organelle rather than just a collection of genes. A notable illustration of the growth of the discipline of bioinformatics, one of the most active study areas in contemporary molecular biology, is the Human Genome Project. The C-value is a measurement of the haploid DNA content. It would seem logical to infer that as organism complexity increases, genome size should grow, reflecting the higher number of genes necessary to support this complexity. Genome size does seem to rise as organismal complexity grows. Thus, the genomes of bacteria, yeast, fruit flies, and humans follow this pattern. Although it may seem counter-intuitive to think that a wheat plant is more complicated than a human, mice, tobacco, and wheat all have considerably larger genomes than humans. Furthermore, given that E. coli has about 4000 genes, it would appear that the tobacco plant genome could encode 4000000 genes, but this is undoubtedly false, even when accounting for the larger size and greater complexity of eukaryotic genes (Vereecke et al., 2020). The C-value paradox is a name that has been given to this peculiarity. Genome complexity also tends to rise with more sophisticated organization, in addition to genome size. Analyzing DNA sample renaturation is one method of investigating complexity. The complementary strands will renature if a DNA duplex is denatured by heating the solution until the strands split. Since sequences that are present in the genome as many copies will renature faster than sequences that are present as single copies only, this trait can be exploited to convey information about the sequence complexity of the DNA in question.

Due to the denatured strands like inverted repeats or palindromes, which fold back on each other to form a hairpin loop shape, certain DNA will first form duplex structures immediately. Foldback DNA is the popular name for this class. In the genome, extremely repetitive regions that appear frequently rank second in terms of speed of re-annealing. The C-value paradox and the eukaryotic genome high sequence complexity raise concerns regarding the structure of genomes. Because of their short genome sizes, viral and bacterial genomes frequently display extremely effective utilization of DNA for encoding their genes. In contrast, just roughly 3% of the DNA in the human genome includes truly coding sequences. The majority of the DNA has regulatory sequences and introns even after they are introduced. The majority of the human genome does not participate in protein coding (Figure 3.7). Although this is possibly the wrong way to think about this DNA, which appears to be redundant, it is commonly referred to as junk DNA.

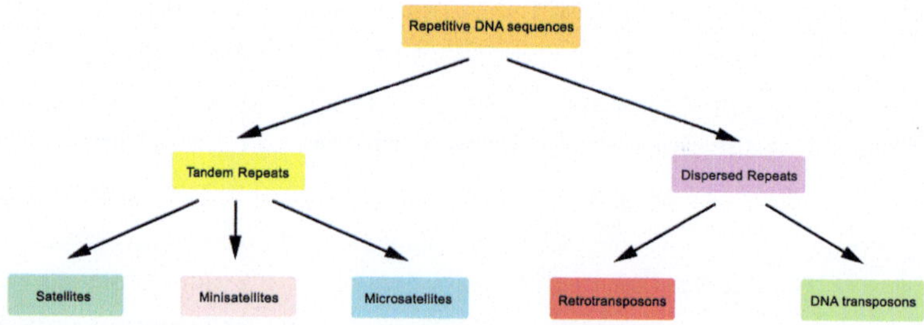

Figure 3.7. Repetitive DNA sequences in eukaryotic genomes. (Source: Paço A, Freitas R, Vieira-da-Silva A. Conversion of DNA Sequences: From a Transposable Element to a Tandem Repeat or to a Gene. *Genes*. 2019; 10(12):1014. https://doi.org/10.3390/genes10121014, Creative Commons License).

Although it is clear that when the complete genome sequence is known, it will be much simpler to identify genes, there are still many instances where the protein products of known genes have unclear biological functions. The nuclei of eukaryotic cells have several chromosomes; however, many eukaryotic genes are single copy genes that are spread out among these chromosomes.

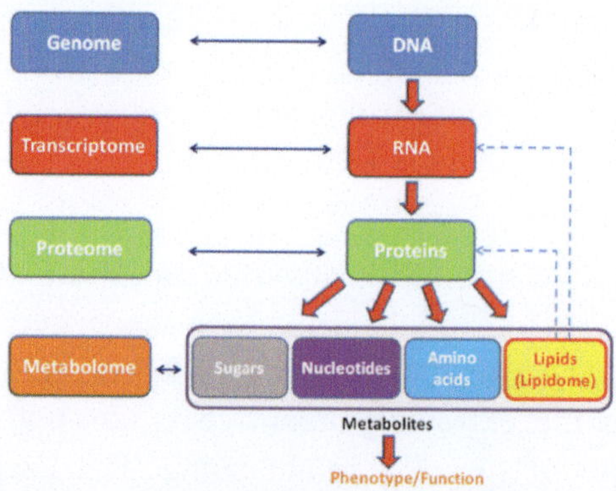

Figure 3.8. Transcriptome. (Source: By Lmaps (talk) - I created this work entirely by myself., CC BY-SA 3.0, https://en.wikipedia.org/w/index.php?curid=21274382, Creative Commons License).

Other genes might be scattered throughout the chromosome or they might belong to multigene families. Genome sequencing has significantly increased our knowledge of how genomes function (Mardis, 2011). Important indications include aspects like gene density, gene size, mRNA size, intergenic distance, and intron/exon sizes. The average size of a coding region in human DNA, according to early studies, is around 1500 base pairs, while the average size of a gene is 10–15 kbp. One gene is present every 40–45 kbp or so, and the intergenic spacing is roughly 25–30 kbp. However, as we have already seen, eukaryotic gene organization can be extremely complicated, making the use of average estimates somewhat misleading. It is evident that as more data is generated by genome sequencing initiatives and we move into what is frequently referred to as the post-genomic age, genomes are now producing a wealth of new knowledge on gene structure and function.

3.10 TRANSCRIPTOME

The population of transcripts present in a cell at any particular time is referred to as the transcriptome (Figure 3.8) (Velculescu et al., 1997). Numerous factors impacting the cell's state will determine this expressed fraction of the genetic data. There will be housekeeping genes for the fundamental upkeep of cell function, but It is also possible that tissue-specific genes are expressed. Developmentally controlled genes might be "on" at that specific time. The transcriptome may therefore be analyzed to get a good idea of what the cell is doing at any given time. The proteome, which reflects the population of proteins in the cell, is a logical extension to the genome and transcriptome. Analysis of the transcriptome will more accurately reflect the transcriptome than the proteome, and it can be used to determine which genes a cell is actively expressing at any particular time. Although certain transcripts might not be translated well and other proteins might survive in the cell even after their transcripts have been withdrawn from circulation, the transcriptome represents the genome. In order to fully comprehend how cells function, many biologists now agree that having a solid understanding of the proteome is essential. The reasoning is that if we fully comprehend how every protein in a cell functions, then we must also fully comprehend the structure and operation of cells. Although the process is not going to be straightforward, as is the case with most biological processes, the next few years will be quite exciting for biologists as the mysteries of gene expression are further exposed. The provision of basic services such as water (including distilled and/or deionized water sources), electrical power, gas, compressed air, vacuum lines, drainage, and so forth. are all examples of general facilities. The majority of the standard services would be offered as part of any laboratory setup and wouldn't pose any unusual challenges or costs.

The exact needs vary based on the sort of work being done, but cell culture and containment facilities are necessary for growing the cell lines and organisms required for the research. The majority of laboratories will need space for bacterial cell growth, as well as autoclaves, incubators (both static and rotational), centrifuges, and protective cabinets. Many people need facilities for the development, containment, and processing of various cells and organisms. It is necessary to use some sort of physical containment to stop organisms from escaping while being manipulated. The type of host and vector being utilized will determine the overall level of containment needed. This combination will provide a level of biological containment because the host is typically rendered inoperable and does not survive outside of the laboratory. The general containment standards may apply to bacterial and mammalian cell culture facilities and are often set by national bodies that control gene manipulation. Therefore, cloning *E. coli* may just need to follow standard microbiological precautions, whereas cloning human genes using viral vectors in mammalian cell lines may need to follow more strict safety protocols. An array of tools is available for the processing and analysis of cells and cell constituents like DNA. The kind of automatic pipette and microcentrifuge tube can be important at the most fundamental level. A researcher will quickly become annoyed if they are attempting to use pipettes that don't function properly or tubes with difficult-to-open caps. Expensive lab tools like automated DNA sequencers and ultracentrifuges may be required, thus care must be taken while selecting them. Many additional pieces of equipment are compact and inexpensive, with researchers possibly favoring a particular brand.

3.11 LAB AND CAREER

The operating costs of a laboratory must be considered in addition to what can be referred to as infrastructure and equipment because they are likely to make up a significant portion of the cost in any given year. Science advances yearly after initial start-up. Research funding is a significant problem for anyone interested in technological advancements, and the level of current knowledge; it is frequently hampered by a lack of funding. Starting a career in science and having the ability to secure substantial research funding are key components of the entire scientific process. Thus, obtaining grants for diverse initiatives may take up a sizable portion of a senior research scientist's work, frequently with little assurance of being able to secure long-term support. There must be a source of nucleic acid, either in the form of DNA or RNA, for any gene-modification procedure. It is crucial that dependable techniques are accessible for separating these components from cells. The sample must be opened to reveal the nucleic acids for further processing, separated from extraneous cell components, and recovered in a pure state. Techniques might range from direct processes with a few steps to more intricate purifications comprising multiple stages. Nowadays, the majority of molecular biology businesses offer kits that make it possible to purify nucleic acids from a variety of sources. Disrupting the starting material, which may be viral, bacterial, plant, or animal, is the initial step in every isolation technique. Cells should be opened with as little force as possible, preferably via enzymatic breakdown of the cell wall. To isolate nucleic acids, cells must be broken; this process should be done carefully to prevent shearing big DNA molecules. There is a risk of shearing or breaking big DNA molecules if more forceful cell disruption techniques are required, which can hinder the formation of recombinant molecules during further processing.

The majority of techniques include a deproteinization stage after cell disruption (Marmur, 1961). One or more extractions employing phenol or phenol/chloroform combinations can accomplish this. The nucleic acids can be precipitated from solution using isopropanol or ethanol. There are several methods that are safer than phenol-based extraction media and do not require the use of phenolic combinations. The RNA in the preparation can be digested using the enzyme ribonuclease (RNase) if a DNA preparation is desired. Once the cell preparations have been split open, a variety of methods can be used to deproteinize them and purify the nucleic acids. Some applications call for highly purified nucleic acid preparations, while others might be able to employ DNA or RNA that has only been partially purified. To further purify eukaryotic mRNAs for cDNA synthesis, affinity chromatography can be used to bind the poly(A) tails of these mRNAs. This significantly increases mRNA enrichment and makes it possible to eliminate the majority of the contaminating DNA, rRNA, and tRNA. Gradient centrifugation is a common method for preparing DNA, especially plasmid DNA (pDNA) (X. Wang et al., 2021). This method involves rapidly spinning a caesium chloride (CsCl) solution containing the DNA preparation in an ultracentrifuge. A density gradient is created over an extended period of, and the pDNA forms a band. Nucleic acid solutions are utilized to make it possible to handle, quantify, and dispense very minute amounts with ease. To obtain a pure preparation of pDNA, the band can be removed and the CsCl can be eliminated using dialysis. Size exclusion chromatography (gel filtering), as an alternative to gradient centrifugation, may be utilized.

3.12 CLONING AND OTHER METHODS

During a cloning experiment, it is frequently required to employ incredibly small amounts of nucleic acid (micro, nano, or picograms). Since handling these quantities directly is obviously impossible, the majority of measurements are performed using aqueous solutions of DNA and RNA. A spectrophotometer can be used to measure the absorbance at 260 nm to determine the concentration of a nucleic acid solution. Monitoring the fluorescence of bound ethidium bromide can also be used to measure the content of DNA in addition to spectrophotometric techniques (Bi, Zhang, Qiao, Sun, & Liu, 2008). When exposed to ultraviolet (UV) light, this stain bonds in between the DNA bases (intercalates) and fluoresces orange. An estimation of the concentration can be made by contrasting the sample's fluorescence with that of a number of standards. When UV-absorbing impurities make spectrophotometric measurements impractical, this approach may be utilized because it can detect as low as 1-5 ng of DNA (Villanueva et al., 2004). Theoretically, any quantity can be dispensed after determining the concentration of a nucleic acid solution by taking the right volume of the solution. This enables the accurate dispensing of nanogram or picogram amounts. The nucleic acid precipitate is recovered by centrifugation and can be processed as needed. Isopropanol and ethanol are the two most often used precipitants, with ethanol being the favored option for most applications. Ethanol causes the nucleic acids to leave the DNA solution when added in a 2:1 volume-to-volume ratio while containing 0.2 M salt. Although it was often believed that low temperatures were required, it appears that 0 degrees Celsius is sufficient. Following precipitation, nucleic acids can be retrieved by centrifugation, which results in the formation of a pellet of nucleic acid material at the tube's bottom. The pellet can be dried, and the nucleic acid can then be redissolved in the proper buffer for the experiments.

Keeping track of the tiny amounts of nucleic acid involved in cloning operations is an important problem. Due to losses, the amount of material often decreases after each step of the process, which makes this problem worse at each stage. One method of tracing the substance is to designate the nucleic acid with some kind of marker so that it can be recognized at each step of the process. Procedures are now well known since radioactive tracers have been employed often in biochemistry and molecular biology for a very long time. Tritium (3H), carbon-14 (14C), sulfur-35 (35S), and phosphorus-32 (32P) are the most frequently used isotopes (Herzstein & Stave, 2016). Tritium and 14C are low-energy emitters, while 32P and 35S are high and medium energy emitters, respectively. Nucleic acids are frequently labeled using radioactive isotopes, despite the fact that these procedures are more dangerous than non-radioactive labeling techniques. Additionally, there are stringent legal regulations for the disposal and storage of radioactive waste. The employment of alternative technologies, such as fluorescent dyes or enzyme-linked labels, has gained popularity recently, in part due to the inherent risks associated with dealing with high-energy isotopes. Although these techniques have benefits for specific applications (such as DNA sequencing), a radioactive label is still frequently the tool of choice for ordinary tracing research. In this instance, radiolabeling is a common term to describe the method. Labeling the nucleic acid with a radioactive molecule, typically a deoxynucleoside triphosphate (dNTP), labeled with 3H or 32P, is one method of tracking DNA and RNA samples. This allows portions of each reaction to be observed/counted in a scintillation counter to determine the amount of nucleic acid present. Calculations are typically used for this, accounting for the sample's level of radioactivity.

3.13 HYBRIDIZATION STUDIES

The creation of extremely radioactive nucleic acid molecules for use in hybridization assays is another use of radiolabelling. These molecules are referred to as radioactive probes, which are highly helpful for recognizing specific DNA or RNA sequences. The primary distinction between labeling for tracing and labeling for probes is specific activity, or the molecule's level of radioactivity. Low specific activity is sufficient for tracing, whereas high specific activity is required for probing. In probe preparation, 32P, a high-energy ß-emitter, serves as the radioactive label. The enzyme polynucleotide kinase is employed in the end-labeling approach to transfer the terminal phosphate group of ATP onto the 5 -hydroxyl terminus of nucleic acid molecules. If the ATP donor is radioactively labeled, only the termini of each molecule become radioactive, resulting in a labeled nucleic acid with a relatively low specific activity. The process of translating a nick made in the phosphodiester backbone of the DNA double helix depends on the enzyme DNA polymerase I. Nicks may develop naturally or may be brought on by a DNase I nuclease concentration that is too low in the reaction mixture. New dNTPs are incorporated into the DNA chain through a strand-replacement reaction that DNA polymerase I catalyzes. Resulting from the use of a radioactive dNTP is a strongly tagged DNA molecule. The term labeling by primer extension refers to a method that uses random oligonucleotides to prime the synthesis of a DNA strand. These oligonucleotides are typically hexadeoxyribonucleotide molecules, which are sequences of six deoxynucleotides. Not all of the radioactive dNTP is integrated into the target sequence in the majority of labeling reactions, and the non-incorporated isotope is often removed before employing the probe.

The oligonucleotide primers anneal to the single-stranded DNA after the target DNA is heated to denature it (Deng & Nickoloff, 1992)requiring only that the target plasmid carry a unique, nonessential restriction site. The procedure employs two mutagenic oligonucleotide primers. One primer contains the desired mutation and the second contains a mutation in any unique, nonessential restriction site. The two primers are annealed to circular single-stranded DNA (produced by heating circular double-stranded DNA. The DNA polymerase's fragment can then synthesize a copy of the template that has been primed with the oligonucleotide's 3 -hydroxyl group. When a tagged dNTP is added, very high specific activity DNA is created. It is frequently desirable to separate the radiolabeled DNA from the unincorporated nucleotides present in the reaction mixture while performing a radiolabelling procedure. Applying a suitable medium to a small-scale gel filtration stage is an easy approach to accomplish this. In a Pasteur pipette, the entire procedure can be completed, with the marked DNA passing through the column ahead of the free nucleotides. The total activity of the DNA, a specific activity, and the degree of incorporation of the isotope can all be determined by collecting and monitoring fractions for radioactivity. Nucleic acid hybridization can also provide information regarding sequence complexity. The simple base-pairing link between complementary sequences has extremely significant ramifications, both for the cell and how it functions as well as for the scientist looking to take advantage of this property. In most cases, a single pure sequence is 32P-labeled and employed as a probe. Before usage, the probe is denatured to allow the strands to base-pair with their complements. Additionally denatured, the DNA that will be probed is often attached to a nylon or nitrocellulose support membrane. To allow the duplexes to form, hybridization is done in a sealed plastic bag or tube at 65-68° C for several hours. Following the removal of the extra probe, the sample can be counted in a scintillation spectrometer to determine the degree of hybridization, or an autoradiograph can be created in which the sample is exposed to X-ray film.

3.14 ELECTROPHORESIS

Since gel electrophoresis (Figure 3.9) is the primary method for directly visualizing nucleic acid fragments, it is essential to the genetic engineer. As a result of the phosphate groups on the phosphodiester backbone of the nucleic acid strands, nucleic acids are polyanionic at neutral pH, which means they carry multiple negative charges. This indicates that when placed in an electric field, the molecules will move toward the positive electrode. Because the DNA molecule negative charges are equally spaced out, the charge/mass ratio is constant, thus mobility is a function of fragment length. The procedure divides the nucleic acid molecules into different sizes using a gel matrix. Gel electrophoresis is one of the most potent molecular biology procedures.

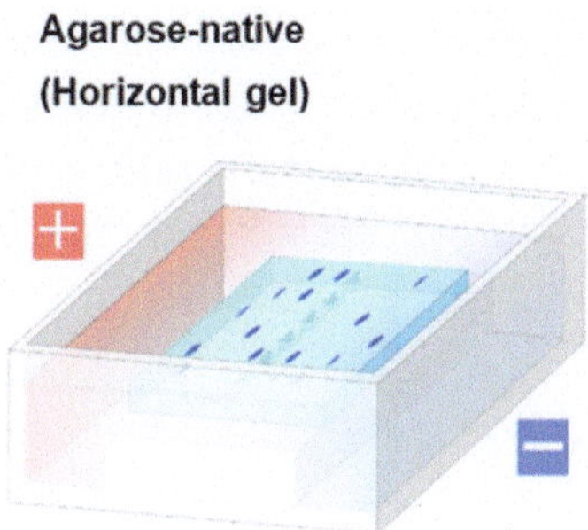

Figure 3.9. Electrophoresis using an agarose gel. (Source: Akuta T, Maruyama T, Sakuma C, Nakagawa M, Tomioka Y, Entzminger K, Fleming JK, Sato R, Shibata T, Kurosawa Y, et al. A New Method to Characterize Conformation-Specific Antibody by a Combination of Agarose Native Gel Electrophoresis and Contact Blotting. *Antibodies*. 2022; 11(2):36. https://doi.org/10.3390/antib11020036, Creative Commons License).

The type of matrix used for electrophoresis has substantial consequences for the degree of separation obtained, which is reliant. In most cases, agarose and polyacrylamide gels are used. Seaweed is used to extract agarose, which is sold as a dry powder that must be melted in a buffer at the right concentration, often between 0.3 and 2.0 percent (w/v). The gel is formed when the agarose cools and settles. The submerged agarose gel electrophoresis (SAGE) method is typically used to run agarose gels in the apparatus. Because the pore size of PAGE is smaller than that of agarose, it is occasionally employed to separate tiny nucleic acid molecules in applications like DNA sequencing. By putting the nucleic acid samples in the gel and applying a potential difference across it, electrophoresis is accomplished. Until a marker dye (bromophenol blue) reaches the end of the gel, this potential difference is preserved. The intercalating dye ethidium bromide is typically used to stain the gel in order to highlight the nucleic acids, which are then visible under ultraviolet light. Nucleic acids appear as orange bands, which are recordable by photography. As migration is inversely proportional to the log10 of the number of base pairs, the data can be used to estimate the sizes of unknown fragments by creating a calibration curve using standards of known size. This is very helpful for the restriction mapping technique.

PAGE is widely used for the analysis of proteins in addition to nucleic acid analysis (Kantardjieff & Rupp, 2003)641 crystallographic Protein Data Bank (PDB. Although the methodology is distinct from that utilized for nucleic acids, the fundamental ideas are the same. SDS-PAGE is a widely used method that denatures multi-subunit proteins and coats the protein molecules with negative charges using the detergent SDS (sodium dodecyl sulfate). In this manner, the charge/mass ratio stabilizes and the protein's inherent charge is concealed. As a result, proteins and DNA molecules can both be classified according to their sizes.

3.15 DNA SEQUENCING

Modern molecular biology places a strong emphasis on the capacity to decipher DNA's base sequence, which offers what might be regarded as the most comprehensive structural knowledge. In the late 1970s, quick methods for sequence analysis were created; today, laboratories all over the world utilize this technology. Automation has revolutionized the fundamental methods in recent years, increasing sequencing productivity to the point that genome sequencing is now feasible. By definition, identifying the bases in a sequential manner that allows for the processive identification of each base in turn is necessary for determining a DNA sequence. The approach utilized must meet the goal of presenting each base in turn in a form acceptable for identification. The detection method must facilitate speedy and precise identification of the bases. DNA fragments must be generated in a form suitable for sequencing.

The synthesis of DNA fragments is direct. DNA sequencing is based on certain simple ideas, but the processes needed to get the desired results are rather more involved. The informatic challenge of determining the location of the segment within the genome is much more challenging. The sequencing methodology is more of a technical than an experimental procedure. Although there are various variations of the fundamental process, the enzymatic method is the technique that is most frequently employed. Whatever the technique, the goal is to produce a series of overlapping segments that end at various bases and have a one nucleotide difference in length. The detection step is the last step in the sequencing process, presuming that the approach has produced a set of nested fragments. The fragments are typically separated in a polyacrylamide gel for this. An autoradiograph is produced by slab gels with radioactively labeled pieces. Fluorescent labeling and continuous electrophoresis are frequently used in automated sequencing processes to separate the fragments, which are recognized as they pass a detector. DNA sequencing is a technique created by Allan Maxam and Walter Gilbert involving using chemicals to break the DNA at certain locations, producing a collection of fragments that differ by one nucleotide (Maxam & Gilbert, 1980). Another technique, created by Fred Sanger and Alan Coulson, uses enzymes to create DNA strands that end in changed nucleotides in order to achieve the same outcome in a different way. The chemical approach and the enzymatic method were two quick sequencing methods created in the 1970s. Modern DNA-sequencing methods are based on the enzymatic approach. The chemical method has now almost entirely been displaced by the enzyme method, while there are some circumstances in which chemical sequencing might offer important data to confirm information generated by the enzymatic method. Radioactive isotopes can be replaced with fluorographic detection techniques. This is crucial for DNA sequencing since it expedites the procedure and makes automation possible.

The size of the sequencing project has a significant impact on how challenging it is to prepare a fragment for sequencing. If the goal is to sequence a portion of a gene that has already been extracted and identified, the process is usually quite simple and only calls for that the fragment is the right size and shape for the sequencing method being used. The challenge is significantly bigger if the goal is to sequence a much larger chunk of DNA, like an entire chromosome. There are two methods for implementing the sequencing technique in this scenario. The first is an ordered sequencing procedure, often known as the clone-by-clone or clone contig method. The approach used to approach large-scale sequencing projects depends on a variety of variables, but ultimately comes down to a choice between the ordered and shotgun approaches. As part of the method, the fragments are monitored, and as the project advances, their relative order is recorded. The order of the fragments is used to put the sequence together. The second strategy is known as shotgun sequencing, in which the fragments are produced and handled at random. The sequence is then stitched together by looking for sequence overlaps on a computer. It is necessary to start with a specific DNA segment. The method can be used on any DNA fragment because it doesn't have to be cloned in a plasmid vector. To create a population of tagged strands for the sequencing processes, the DNA is radiolabelled with 32P at the 5' ends of each strand, and the strands are then denatured, sorted, and purified. The nucleotides in the DNA strand are then chemically altered in the following step. This is accomplished through a sequence of four to five distinct reactions, each of which is carried out under specified conditions so that, on average, just one alteration is introduced into each copy of the DNA molecule. The strands are then split at these locations by the chemical piperidine after the modified bases have been separated from their sugar groups. According to the hypothesis, this process will result in a collection of nested fragments due to the many molecules involved and the variety of reactions.

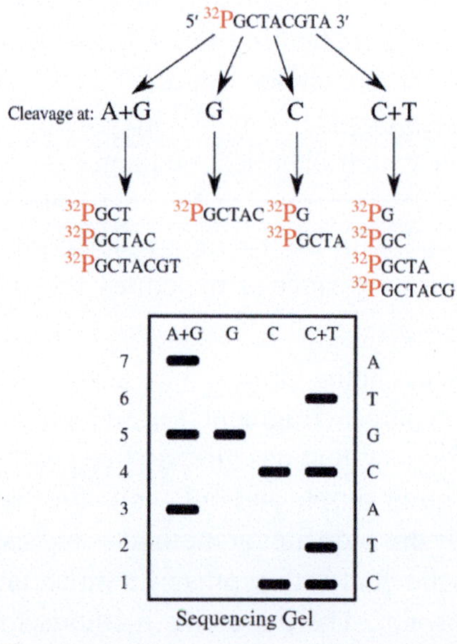

Figure 3.10. Maxam and Gilbert gene sequencing. (Source: By w:User:Several Times (File:Maxam gilbert sequencing.png)Shakiestone (vectorization)Incnis Mrsi (tweaks) - This is a retouched picture, which means that it has been digitally altered from its original version. Modifications: English, formatting. The original can be viewed here: Maxam-Gilbert sequencing.svg: . Modifications made by Incnis Mrsi., CC BY-SA 3.0, https://commons.wikimedia.org/w/index.php?curid=27276995, Creative Commons License).

Though the process used by Sanger-Coulson and Maxam and Gilbert (Figure 3.10) is completely distinct from one another, the end result is comparable to that obtained by the chemical technique. In this instance, DNA polymerase creates a copy of the DNA that will be sequenced. Single-stranded DNA serves as the reaction's template, and a primer is required to give DNA polymerase the 3 terminal it needs to start synthesizing the copy. The insertion of a modified dNTP in each reaction results in the formation of nested fragments. These dNTPs lack the hydroxyl group that deoxyribose needs in order to proceed with chain elongation at position 3 . Dideoxynucleoside triphosphates are the name given to these modified dNTPs (ddNTPs). A set of four reactions, each including the four regular dNTPs, includes the four ddNTPs (A, G, T, and C forms). The dideoxy form will only seldom be absorbed into the developing DNA chain because of the dideoxy form's concentration. Therefore, each reaction generates a set of fragments that end at a certain nucleotide, and the four reactions combined to generate a set of nested fragments. By introducing a radioactive dNTP in the reaction mixture, the DNA chain is marked. When creating fragments for dideoxy sequencing, which is more difficult than chemical sequencing, various vectors are typically subcloned. There are currently many plasmid vectors available, some of which can be employed straight away for DNA sequencing research. Another approach is to clone the DNA into a vehicle that makes single-stranded DNA during infection, like the bacteriophage M13. This gives the sequencing processes a sufficient substrate.

Further, PAGE is used to separate the DNA fragments generated during sequencing processes (Ansorge, 2009). A single gel system is employed for the typical lab technique. In order to lessen the impacts of DNA secondary structure, the gels typically include 6-20 percent polyacrylamide and 7 M urea, which denatures the DNA. This is significant since it allows for the separation of pieces that only differ in length by one base. The gels are extremely thin (0.5 mm or less), and because they are operated at high power levels, they can reach temperatures of 60–70 °C. Additionally, it keeps the conditions conducive to denaturing. To maximize the quantity of sequence information acquired, two sets of samples may occasionally be placed onto the same gel at different times. The gel is withdrawn from the apparatus after it has been run and may be dried onto a paper sheet to make handling easier. An X-ray film is then used to capture it. The radioactive label emissions sensitize the silver grains, causing them to turn black when the film is developed and fixed. An autoradiograph is the result. It is simple to read the autoradiograph; the sequence is read from the smallest fragment up. Sequences of up to several hundred bases may be read from a single gel using this technique. A computer is then used to compile and analyze the sequence data, doing tasks including translating the data into amino acid sequences and finding restriction sites, sequence homology regions, and other structural patterns like promoters and regulatory regions. It was necessary for sequencing technology to automate many steps of the process. In order to complete genome-sequencing projects in a reasonable amount of time, these methods must be improved. A competent lab worker or scientist might be able to sequence a few hundred bases every day, but this would not be sufficient to distinguish between genome and gene sequences. It was necessary to advance technology. This was accomplished through advancements in sample handling and preparation, with robotic processing permitting high throughput. Similar to this, the scaling up of the sequence-determination portion of the process was made possible by the automation of the sequencing operations using capillary electrophoresis techniques.

In addition to the difficulties in determining the sequence, another difficulty was in developing the necessary processing capacity to handle the enormous volumes of data produced by the newly enhanced technology used in genome sequencing. In fact, it could be argued that the data analysis aspect of the process is the most important one because, without the ability to query sequence data, the sequence effectively remains mute. One of the most significant categories of enzymes for the modification of DNA is the restriction enzymes, which cut DNA at predetermined locations. These enzymes are present in bacterial cells and are a component of the restriction-modification system, which is a defense mechanism. In this system, any exogenous DNA that enters the cell is hydrolyzed by the restriction enzyme. Restriction enzymes hydrolyze exogenous DNA that has not been methylated by the host modification enzyme, serving as a protection strategy for bacteria (Dyachenko, Tarlachkov, Marinitch, Shevchuk, & Buryanov, 2014)Tarlachkov, Marinitch, Shevchuk, & Buryanov, 2014. By methylating specific bases in the restriction enzyme's recognition sequence, the system's modification enzyme alters the host DNA and inhibits the enzyme from cutting the DNA. There are three types of restriction enzymes (I, II, or III). The majority of enzymes in use today are type II enzymes, which have the simplest modes of action. These enzymes are nucleases, and because they start the breakdown of a DNA strand at one end rather than the other, they are also referred to as endonucleases. Although they are frequently referred to as simply restriction enzymes, these enzymes are actually type II restriction endonucleases. They can be compared to molecular scissors.

Depending on the bacterial strain involved and the presence or lack of extrachromosomal components, additional descriptors might be included.

EcoRI and BamHI are two commonly used enzymes from the bacterium. The specificity of restriction endonucleases is what makes them useful. Each enzyme recognises a particular pattern of DNA bases, with the most frequent patterns being four, five, or six base pairs long. Given that DNA has four bases and that different restriction enzymes produce DNA fragments of varying length, the size of the fragment is correlated with the frequency with which the recognition sequence occurs. The predicted frequency of any given sequence can be estimated as 4n, where n is the length of the recognition sequence, under the assumption that the bases are distributed randomly. According to this, tetranucleotide sites, pentanucleotide sites, and hexanucleotide sites should appear every 256, 1024, and 4096 base pairs, respectively (Klump, 1987)pentamer-, and hexamersequences. It is basically assumed that each base pair within these sequences has to be recognized by directionally unique bidentate hydrogen bonds either within the plane of the base pair or by bridging the appropriate H-bond donor/acceptor groups of the neighbouring bases of the same strand. Thus sequence specificity is mediated by twelve (eight. Although there is, as one might anticipate, a wide range from these values, in general, the resulting fragment lengths will be close to the estimated value. A reasonable amount of restriction enzyme is added to the target DNA in a buffer solution, and the reaction is incubated at the ideal temperature (37°C) for a suitable period of time. Restriction enzymes are fairly easy to use. One unit of an enzyme's activity is equal to how much of the enzyme can break one microgram of DNA in an hour at 37°C. There are some situations where different combinations of enzyme concentration and incubation time may be utilized to achieve only partial digestion, even though most tests call for complete digestion of the target DNA.

The recognition sequence and the position of the cutting site within this sequence determine the type of DNA fragment that a certain enzyme creates. The ability of restriction enzymes to produce cohesive or sticky ends that may be utilized to connect DNA from two separate sources together to create recombinant DNA molecules is a particularly important property of these enzymes. The sort of ends that the cut fragment possesses will depend on the enzyme's actual cutting location, which is crucial for further DNA manipulation.

Chapter

4 GENETIC ANALYSIS

CONTENTS

4.1 INTRODUCTION

Genetics is a subfield of information science and geneticists work to understand the rules that govern the transmission of genetic information at three different levels: from parent to child within families, from DNA to gene action within and across cells, and across many generations within populations of animals and other species. These three branches of genetics are transmission genetics, molecular developmental genetics, and population-evolutionary genetics.

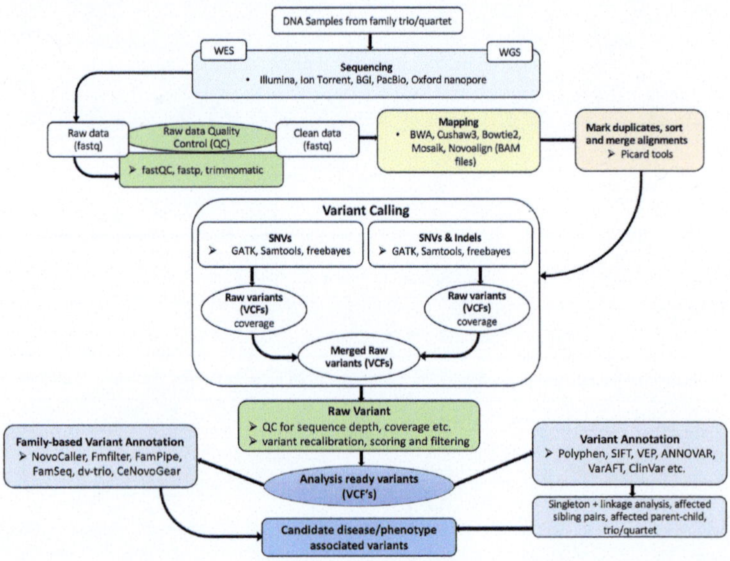

Figure 4.1. Genetic inheritance and latest techniques. (Source: Kanzi AM, San JE, Chimukangara B, Wilkinson E, Fish M, Ramsuran V and de Oliveira T (2020) Next Generation Sequencing and Bioinformatics Analysis of Family Genetic Inheritance. Front. Genet. 11:544162. doi: 10.3389/fgene.2020.544162, Creative Commons License).

Genetics became a recognized field for the first time just over a century ago. Genetics has since profoundly changed our understanding of life at all scales, from the level of a single cell to a community of animals evolving over millions of years. In the 1800s, biologists, animal breeders, and horticulturists tried to explain why parents and children are alike. It was generally believed at the time that inheritance worked similarly to mixing liquids like paints, which is known as the theory of inheritance. It makes sense that a child with one tall parent and one short parent will grow up to be a medium height because pink is created when red and white are mixed.

It was clear that the concept of blending had exceptions, such as tall children born to parents of average height.

4.2 MENDELIAN EXPERIMENTS

While the advantages and drawbacks of blending theory were being addressed, an Austrian monk by the name of Gregor Mendel was working to understand the laws underlying the transmission of traits from parent to offspring after hybridization among numerous species of pea plants (Schwarzbach et al., 2014).

He experimented with breeding a purple-flowered pea variety with a white-flowered variety. Mendel reported that the first hybrid generation of this cross exhibited purple flowers in all of the offspring, much like one of the parents. There was no blending. By self-pollinating the first hybrid plants, Mendel created a second generation of hybrid plants. He observed that 224 of the 929 plants had white flowers, and 705 had purple flowers. He observed that there were typically three plants with purple flowers for every plant with white flowers. A uniform population of first-generation hybrid plants with light purple flowers is required. Therefore, Mendel proposed that the factors that affect features behave like particles rather than fluids, and that these particles do not mix but rather are passed on in their totality from one generation to the next. Genes are the modern name for Mendel's particles or factors.

Mendel discovered that each pea plant has two copies of the gene that controls flower color in each cell (somatic cells). However, when the plant produces sex cells, or gametes, only one copy of the gene penetrates the reproductive gametes. In the event that the egg and sperm combine to create a new individual, the flower color gene will then be present twice in each cell of the plant body. Mendel had some additional viewpoints. According to his theory, there are two gene variants, or alleles, that change the color of flowers: one is responsible for white flowers and the other for purple flowers. His hypothesis states that because the purple allele of the flower color gene is dominant to the white allele, a plant with one purple allele and one white allele will produce purple flowers. Only plants with two white alleles would have white flowers. Mendel came to the two conclusions that genes behaved like particles or factors that do not blend together and that one allele is dominant to the other in order to explain why there was no blending in the first generation of hybrids and why white-flowered plants reappeared in the second generation of hybrids with a 3:1 ratio of purple-to white-flowered plants. Mendel made a wise choice in the organism and characteristics to study. He found that a single gene controlled every trait. It would not have been as easy for him to understand the laws of heredity if he had not possessed numerous traits that are controlled by numerous genes. Mendel was a thorough recorder of his experiments, and he was a meticulous observer as well. Not to mention, Mendel was a visionary thinker who was able to look well beyond the ideas of his time. Bateson had a copy of Mendel's 1866 book on the cross-pollination of plant species. After citing Mendel's original study, Bateson had learned that biologists in Germany, the Netherlands, and Austria had independently verified Mendel's 3: 1 ratio. These three individuals had rediscovered Mendel's laws of inheritance. He understood that the inheritance conundrum had been resolved. He soon had a strong belief in Mendel's inheritance theories. A few years later, in 1905, Bateson gave the field of study of inheritance the name genetics (Richmond, 2006, pp. 1895–1910). The genetics revolution had begun.

After Mendel's inheritance rules were discovered in 1900, a burst of original concepts followed. Mendelism evolved to become the organizing principle for a substantial portion of biology. There were many new questions to be asked concerning inheritance. In 1910, Thomas H. Morgan of Columbia University in New York demonstrated that Mendel's genes can be found on chromosomes, thus validating the chromosome theory of inheritance. It wasn't a new idea. Kansas farmer and American army physician Walter Sutton proposed the chromosomal theory of heredity for the first time in 1903. Later, during World War I, Sutton worked as a surgeon for the American military. Theodor Boveri, a German biologist, independently proposed it at the same time. Although it was a compelling theory, there were no supporting experimental findings. This changed in 1910 when Morgan utilized the fruit fly as his experimental organism and Mendelian genetics to show how the chromosomal hypothesis of heredity works. While simple qualities like flower color might be directly detected as having a 3: 1 segregation ratio, many traits in second-generation hybrids exhibit a continuous range of values without such simple ratios as 3: 1. In 1918, British geneticist and statistician Ronald Fisher discovered a solution to the Mendelian gene theory of the inheritance of variable human traits like height.

4.3 THE EXAMPLE OF LACTASE

Human communities survived on food derived from the environment between 10,000 and 12,000 years ago through wild animal hunts and the harvest of fruits and vegetables. At the time, there were no human populations consuming dairy products. Techniques for milking cows had not yet been devised, and cattle had not yet been domesticated. Mother's milk served as an infant's main source of sustenance, but as they grew older, the lactase gene which aids in the digestion of lactose was turned off. In pre-agricultural societies, the lactase enzyme was no longer, hence the lactase gene contained a regulatory element that turned it off in late childhood. With the introduction of agriculture, cattle began to be domesticated. It is conceivable that the first farmers started off by raising cattle for meat. Following the development of milking, milk became a new source of sustenance. But there was a problem. Children in these primitive societies could metabolize milk sugar even while adults couldn't. Adults could drink milk, but because they couldn't digest the lactose, they would feel bloated, and lightheaded. Adults with lactose intolerance have these symptoms following milk consumption. Importantly, because they were unable to digest milk sugar, they were not utilizing this source of nutrition. In ancient communities when food was periodically scarce, making the best use of all available food sources might mean the difference between life and death. Adults, however, are unable to metabolize milk sugar or lactose since they do not have the lactase gene. The many forms of the trait are known as phenotypes.

4.4 MUTATION AND EVOLUTION

Compared to the wild type, mutants are rare. We are aware that they diverge from wild types through the process of mutation, which results in a heritable change in the DNA of a gene. The gene's changed form is also referred to as a mutation. Mutations are not always detrimental to an organism, despite the fact that they frequently have no noticeable effects. Rarely, they might even be useful. It can typically be asserted that errors in DNA processing by cells are the origin of mutations, although there is a ton of information on the causes of mutation. The majority of natural populations display polymorphisms, which are defined as the coexistence of two or more frequently occurring phenotypes of a biological trait (Skulason & Smith, 1995)including morphological, behavioral and life history characters. Research on a number of unrelated taxa suggests that resource polymorphisms may be underestimated as a diversifying force and potentially play important roles in population divergence and initial steps in speciation. In an ecological context, they are important in resource partitioning and reducing intraspecific competition. Recent research suggests that the mechanisms maintaining these polymorphisms may be similar in diverse taxa, that phenotypic plasticity is important, and that some are under simple genetic control.","container-title":"Trends in Ecology & Evolution","DOI":"10.1016/S0169-5347(00. The occurrence of plants with red and orange fruit in a population of wild raspberries serves as an illustration of this. Although polymorphisms can be employed in genetic analysis, one problem is that they frequently do not include the precise feature that the researcher is interested in. Mutants are far more useful since they allow the researcher to concentrate on any property. The functional analysis via gene discovery comprises compiling mutants that affect the assayed biological characteristic, hybrid mutant animals with wild-type individuals to see if the offspring show wild-to-mutant ratios that are suggestive of inheritance from a single gene, ascertain the gene's molecular roles, deduce the genetic interactions that result in the desired phenotype. A quest to gather mutants with the biological function under study altered or missing is the first step in the gene discovery process. Mutants may not be prevalent on their own, but there are techniques to induce their recovery. Exposing the organism to radiation or chemicals that induce mutations is a popular practice. After treatment, visual screening of a sizable population to discover if any of them happen to be mutants by accident is the fastest technique to identify mutants. Various selection methods can also be devised to improve the desired types.

One can hope that each mutant in a group of mutants that affect the target trait represents a cut in the gene responsible for the desired property. Because not all mutants are caused by lesions inside a single gene, each mutant must first be assessed to determine if it is, in fact, caused by a single-gene mutation. Mutant people are mated with their wild-type counterparts, and the first and second generations of progeny are then examined in order to test for single-gene inheritance. As an example, a mutant plant with white flowers would be crossed with a red-flowered wild type. The characteristic ratios of plants with red flowers to those with white flowers in each generation will reveal if a single gene controls flower color. If this is the case, it follows that the wild type would be encoded by the gene's wild-type form, but the mutant would be encoded by a version of the same gene in which a mutation event had modified the DNA sequence in some way. The identification of a collection of flower-color genes would result from the investigation of other mutations that change flower color. Since the biological sample in question is dissected using mutations rather than a knife to reveal its underlying genetic program, this technique of utilizing mutants is frequently referred to as genetic dissection. Each mutation can indicate a different gene that affects that trait. Once a collection of crucial genes has been discovered in this manner, various molecular methods can then be used to determine the activity of each of the essential genes.

This approach to discovering novel genes, which is also known as forward genetics, begins with random single-gene mutations and concludes with their DNA sequence and biochemical function in order to understand biological function. The arrangement of genes on chromosomes is diagrammatically represented by a unidimensional chromosomal map, which displays gene sites, also known as loci, and the distances between the loci depending on some type of scale. Although they were put together in quite different ways, the two primary categories of chromosomal maps used in genetics today are complementary. Recombination-based maps pinpoint the locations of genes found through mutant phenotypes exhibiting single-gene inheritance. On physical maps, the genes are shown as segments that are arranged along the extensive DNA molecule that makes up a chromosome. These maps show different aspects of the genome, but when combined, they can be used to understand how a gene's molecular function impacts phenotype. Using a method known as linkage analysis, chromosome recombination maps are frequently constructed together. If two loci are on the same chromosome, then any homologous alleles are physically connected (linked) by DNA when two genes are said to be linked in genetics.

Early in the twentieth century, William Bateson and R. C. Punnett, who gave the Punnett square its name, was studying the inheritance of two genes in sweet peas (Darden, 1977). The ratio of 9: 3: 3: 1 that the principle of independent assortment predicts did not exist in a dihybrid F1. In fact, Bateson and Punnett noticed that certain allele combinations showed up more frequently than they should have, almost as if they were connected physically in some manner.

A similar deviation from Mendel's second law was later found by Thomas Hunt Morgan (Figure 4.2) while studying two autosomal genes in the Drosophila. Morgan proposed linkage as a theory to account for the occurrence of apparent allele linakge (Greenwood et al., 2000). The vg, vestigial, and vg +, wild-type genes regulated wing length, while the pr, purple and pr +, red genes affected eye color. Vestigial wings are substantially smaller than wild-type wings. Wildtype alleles of both genes predominate. It is important to take note of Morgan's use of the test cross. Because of this, the analyst can concentrate on meiosis in the dihybrid parent while mostly disregarding meiosis in the other parent.

Figure 4.2. Thomas Hunt Morgan. (Source: By Carl Rudolph Gist - The Big T (yearbook of the California Institute of Technology), 1931, Public Domain, https://commons.wikimedia.org/w/index.php?curid=77789798, Creative Commons License).

On the other hand, there are two sets of meiosis to consider when analyzing children from an F1: one in the male parent and the other in the female. The linkage hypothesis explains why allele combinations from the paternal generations stay together by describing how the chromosomal tract that exists between the genes connects them. During the process known as crossing over, Morgan postulated that homologous chromosomes occasionally break. When duplicated homologous chromosomes pair with one another or, in genetic terminology, when the two dyads come together as a bivalent, a chiasma, also known as chiasmata, usually forms between two non-sister chromatids. Morgan thought that the aesthetic appeal of the chiasmata supported the notion of crossing over. The essential method for mapping genes using recombinant frequencies was developed by one of Morgan's students. Morgan discovered that the proportion of recombinant progeny varied greatly depending on which connected genes were being researched as he studied more and more linked genes. This variance in recombinant frequency, according to Morgan, may possibly reveal the real distances between genes on chromosomes.

4.5 TEST CROSS

Morgan assigned Alfred Sturtevant, an undergraduate who later became well-known as a geneticist, the duty of quantifying the process. Morgan asked Sturtevant to interpret the data on gene crossing between numerous related genes. Sturtevant developed a method for mapping genes that is still used today. Sturtevant's logic is illustrated by Morgan's testcross results with the pr and vg genes, from which he calculated a recombinant frequency of 10.7 percent. Sturtevant asserts that we can use this recombinant percentage as a numerical measure of the linear distance between two genes on a genetic map, or linkage map as it is also known. This entire idea is based on reasoning. Think about two specific genes that are separated by a certain amount. Think about randomly moving along the matched homologs. When non-sister chromatids infrequently cross over in the region of the chromosome between these genes during different meiosis, recombinants are created. No recombinants form from these meiotic divisions; these genes do not cross over in other meiotic divisions. According to Sturtevant's theory, the fraction of recombinants produced grows as the likelihood of gene crossovers in the region between the connected genes increases with increasing distance. As a result, by estimating the frequency of recombinants, we may determine the map distance between the genes. Sturtevant really defined one genetic map unit (m.u.) as the distance between genes for which 1 in 100 meiotic products are recombinant. For instance, Morgan's 10.7% recombinant frequency (RF), which is equal to 10.7% of the unit, is 10.7%. A map unit is occasionally referred to as a centimorgan in honor of Thomas Hunt Morgan (cM). Sturtevant predicted that if genes A and B are separated by 5 map units (5 m.u.) and genes A and C are separated by 3 m.u., then the distance between genes B and C should be either 8 or 2 map units on a linear map. Sturtevant found that his prediction had come true. In other words, his work provided strong evidence that genes are organized in a linear order, which would make map distances additive. Since we now understand from molecular research that a chromosome is a single DNA molecule with the genes arrayed along it, it comes as no surprise to us that recombination-based maps are linear since they reflect a linear array of genes.

4.6 GENOMICS

The amount of repetitive DNA found in most genomes is one finding of molecular genomic research (Houck, Rinehart, & Schmid, 1979)the Alu family is shown to be distributed over a minimum range of 30% to 60% of the genome. (The breadth of this range reflects the presence of inverted repeated sequences which, in part, include the Alu family.. Furthermore, there are many different types of repetitive DNA. Despite the fact that it is unknown where these repetitions originate, the fact that various individuals frequently have differing numbers of copies makes them useful. Therefore, these repetitions are known as simple sequence length polymorphisms (SSLPs). They are also known as variable number tandem repeats or VNTRs. An SSLP gene has been found to have up to 15 alleles, which is normal for SSLPs. As a result, it is occasionally possible to trace the lineage of 4 alleles, 2 from each parent. Both minisatellite and microsatellite markers are SSLPs that are useful for studying and mapping the human genome. A minisatellite marker is employed based on changes in the number of tandem repetitions of a repeating unit between 15 and 100 nucleotides long. In humans, its overall size falls between 1 and 5 kb. The genome has a large number of minisatellite loci with identical repeating units but different repeat counts. A microsatellite marker is built on an even shorter sequence, which is often composed of just a few nucleotides, such as a dinucleotide. This sequence may contain different numbers of tandem repetitions. Simple sequence length polymorphisms are found by making use of the fact that homologous regions with different numbers of tandem repetitions would have different lengths. A typical method for identifying these differences is to use flanking areas as primers in a PCR analysis. PCR is used to replicate the DNA sequences until there is enough for further analysis (Wilson, DiZinno, Polanskey, Replogle, & Budowle, 1995)such as: chemical contaminant effects on DNA from blood and semen and the effect of typing DNA extracted from body fluid samples deposited on various substrates. A second set of experiments was performed strictly on human hair shafts. These studies included typing mtDNA from hairs that were: (1. The various lengths of the generated PCR products can be determined using the varying mobilities of the sequences on an electrophoretic gel. DNA fingerprints are a common term used to describe the patterns produced on the gel in the case of minisatellites.

4.7 PHENOTYPE MAPPING WITH SNPS

When we locate a gene, whose phenotypes are regulated by a single nucleotide alteration, we effectively map an SNP. The same technique used to map gene loci can be used to map SNPs without a distinct phenotype. Take into account an individual who has an AT at position 5658 on one chromosome and a GC base pair at the same location on the other. Such a person has the molecular heterozygosity AT/GC at that DNA site. This information is useful for mapping since molecular heterozygotes (AT/GC) can be mapped similarly to phenotypic heterozygotes (A/a). Similar to how heterozygous phenotypic alleles are placed into a chromosomal map, molecular heterozygote locations can be entered using recombination frequency analysis. Molecular markers can be used to identify a gene of interest. They are very useful for letting you know when you are close to your target.

Consider a specific genetic scenario in which we need to locate a disease gene in mice in order to determine its DNA sequence. Every time, we combine a carrier of the disease gene with a carrier of a certain molecular marker whose position on the map is previously known. The next step is to perform recombination analysis to determine whether the target gene is related to any of the molecular markers that were detected using PCR on the parents and progeny at known map coordinates. The results of these crosses might indicate that one of these markers, designated M and located 2 M.U. from the disease gene, is connected to the disease. The method has given us an approximate estimate of the location of the disease gene on the chromosome. The original identification of the cystic fibrosis gene's location was made possible by its relationship with molecular markers known to be on chromosome 7. The gene was isolated and sequenced as a result of this finding, and it was later discovered that the gene encodes a protein that is now known as the cystic fibrosis transmembrane conductance regulator (CFTR) (Tsui, 1995)significant progess has been made in elucidating the mutational basis for this severe, autosomal recessive disease. Such advances have been of major importance in furthering our understanding of the basic defect in CF. Studies of the protein product of the CF gene, referred to as the CF transmembrane conductance regulator (CFTR. The Huntington's disease gene, which produces the huntingtin muscle protein, was also discovered using this technique.

4.8 CENTROMERE MAPPING

Centromeres are DNA sections rather than genes, yet geneticists are very interested in them because they are crucial for the reproduction of living things. Because they lack the heterozygosity required to act as markers, most eukaryotes' centromere loci cannot be mapped via recombination analysis. The fungi that produce linear tetrads, however, allow for the mapping of centromeres. We can use the fungus Neurospora as an example. In haploid fungi like Neurospora, the haploid nuclei from each parent combine to form a temporary diploid that undergoes meiotic divisions down the long axis of the ascus. Along with the four meiotic products, these eight ascospores also contain post-meiotic mitosis products (a tetrad). Centromere mapping assesses a gene locus and determines how far it is from the centromere. The strategy is based on the finding that when a gene and its centromere cross paths during meiosis, a certain arrangement of alleles would appear in a linear tetrad or octad. These loci are localized using the frequency of recombinants during meiosis. The mapping unit is recombinant frequency since it is thought to be proportional to the distance between two loci on a chromosome. This recombination-based approach has been used to map genes with known mutant phenotypes for almost a century. Molecular heterozygosity hotspots that are unrelated to mutant phenotypes can also be found on such recombination maps. These molecular markers are used to direct toward a gene of biological significance after being mapped via recombination like any heterozygous site. Recombination maps are thought to depict the location of genes on chromosomes, although as was previously said, these maps are only models. On the other hand, physical maps are the closest thing to a real genomic map that science has to provide.

4.9 MAPPING

Physical maps are basically representations of genomic DNA, which is a very lengthy sequence of DNA nucleotides (Mozo et al., 1999)detailed YAC, BAC, P1 and cosmid-based physical maps (that is, representations of genomic regions as sets of overlapping clones of corresponding libraries. It displays the position of the genes as well as their size, size variation, sequencing, and other relevant markers. On a physical map, distance is measured in terms of the number of DNA bases; the kilobase is the most useful measurement. The sequence of a DNA molecule is obtained by sequencing a large number of minuscule genomic fragments and then assembling them into a single, complete sequence. A computer that has been programmed to look for segments that resemble genes and are distinguished by specific base sequences, including well-known signal sequences for the start and halt of transcription, scans the sequence next. A gene is compared to a public database of other sequenced genes whose functions have been determined in other organisms once it has been identified by computer tools. A hit, or when the sequence significantly resembles that of a gene with a known function in a different species, occurs frequently. In other situations, the two genes' actions might also be similar. Sequence similarity, which is frequently close to 100%, is explained by the inheritance of the gene from a common ancestor and the general conservation of functional sequences over evolutionary time. The system discovered additional genes with no sequence similarity to any known-function gene. They could therefore be described as genes in search of a function. Of course, the person looking for the function is the researcher, not the gene. Sequencing a population can also be used to find sites of molecular heterozygosity, which act as orientation markers on the physical map similar to how they do in recombination maps.

In a dihybrid test cross of Drosophila, Thomas Hunt Morgan found a deviation from Mendel's law of independent assortment. The two genes, in accordance with his theory, were located on the same homologous pair of chromosomes. The term for this connection is linkage. However, linkage does not explain how the recombinant (non-parental) gene pairings arise. Linkage explains why the parental gene pairings continue. Morgan postulated that chromosomal fragments might physically exchange during meiosis through a mechanism now known as crossing over. At the four-chromatid stage of meiosis, crossing over happens as a result of the actual division and recombination of chromosomal fragments. Meiotic recombination can take one of two main forms. Through Mendelian independent assortment, a 50% recombination frequency is attained. After crossing over, recombinant frequencies are frequently less than 50%. A wide range of recombinant frequency values were discovered when Morgan looked at more related genes, and he questioned whether these values really reflected the actual distances between the genes on a chromosome. Alfred Sturtevant, a mentor of Morgan's, developed a technique for calculating the distance between genes on a linkage map based on the RF. The easiest way to assess RF is using a test cross of a dihybrid or trihybrid. The loci of the examined genes can be seen on a chromosomal map using RF values represented as percentages as the map units. The frequency of second-division segregation in ascomycete fungi can also be used to map the location of the centromeres. Single nucleotide polymorphisms are single nucleotide differences in DNA sequences. Single sequence length polymorphisms are variations in the number of repeating units. SNPs and SSLPs are useful molecular markers for gene mapping. Although the basic test for linkage is a deviation from independent assortment, this deviation may not be evident in a testcross, necessitating the application of a statistical test.

An underestimation of map distance based on RF may result from the nonrecombinant chromatids generated by some repetitive crossings. Any mapping function that is organism-specific adjusts for this predisposition. The Perkins formula is utilized in the same manner in fungal tetrad analysis. In genetics generally, a recombination-based map of loci causing mutant phenotypes is used in conjunction with a physical map like the whole DNA sequence, which depicts all the gene-like sequences. Knowing a gene's position in both maps allows one to combine cellular function with a gene's influence on phenotype. The crossing-over mechanism is thought to start at a double-strand break in one participating chromatid. When a single strand enters the double helix of the other chromatid that is involved, heteroduplex DNA is produced. Gaps are filled during the polymerization process. This structure's molecular resolution becomes a full double-stranded crossing at the DNA level. DNA technology is responsible for the rapid genetic advancements being made in all model organisms. Examples include the recently disclosed whole genome sequences of humans and chimpanzees, which have received extensive media coverage, as well as the frequent use of DNA-based forensic investigation in television and cinema. In reality, the genomes of hundreds of different species have been sequenced due to advancements in technology. All of these extraordinary results, whether they are seen in humans, fish, insects, plants, or fungi, are the result of the use of techniques that enable small pieces of DNA to be extracted, moved from cell to cell, and amplified into pure quantities. The intricate systems that allow for these DNA alterations in any organism are produced by bacteria and their viruses. So, the advancement of modern genetics to its current state of understanding was entirely due to the discovery of bacterial genetics.

4.10 BACTERIAL GENETICS

The goal of bacterial genetics has never been to make eukaryotic molecular genetics simpler (J. R. Brown & Doolittle, 1997)determining the phylogenetic relationships among the contemporary domains of life, the Archaea (archaebacteria. It is impossible to exaggerate the significance of microorganisms to biology. They are the most prevalent species on our planet. They help ecosystems recycle nutrients like nitrogen, sulfur, and carbon. Some of these are pathogens that affect people, animals, and plants. In our mouths and intestines, several species survive. Many bacterial species are useful for the industrial production of a wide range of chemical molecules. In order to understand their biological role, researchers have studied the genetics of bacteria for the same reason they studied the genetics of multicellular animals. Blue-green algae and bacteria both belong to the prokaryote class of life. Because their DNA is not housed within a nucleus that is membrane-bounded, prokaryotes are distinguishable from other organisms. Similar to higher species, bacteria have chromosomes, which are lengthy DNA-based gene sequences. However, a number of factors give their genetic material's organization a unique trait. The majority of bacteria have a single, closed-circle-shaped double-stranded DNA molecule that makes up their entire genome. Naturally existing bacteria usually contain plasmids (Figure 4.3), which are extrachromosal DNA components.

Similar to DNA, the majority of plasmids are much smaller than the entire bacterial genome. Bacteriophages, or phages in general, are specific viruses that can cause bacteria to become infected. The animals that we have so far researched differ greatly from phages and other viruses in terms of structure and function. Viruses and other biological organisms have several traits in common. They may have DNA or RNA as their genetic material, which is a single, condensed chromosome. However, the majority of scientists believe viruses to be non-living because they are not cells and do not have their own metabolism. In order to study the genomes of viruses, they must be created in the cells of their host organisms.

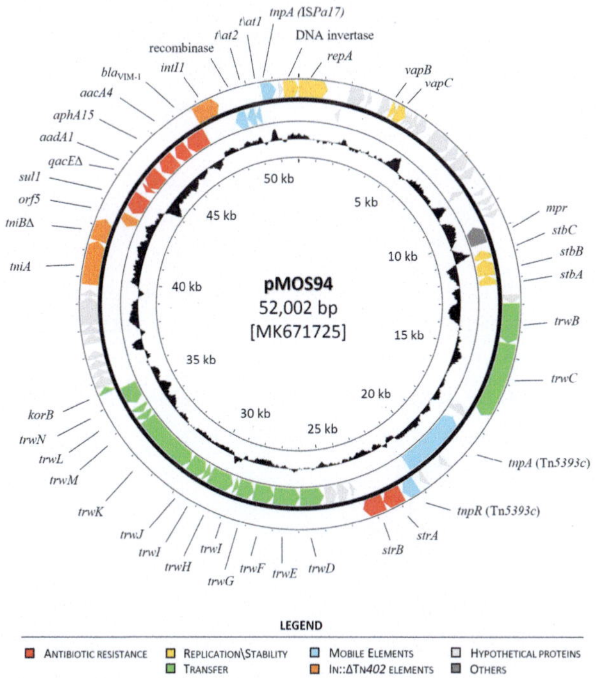

Figure 4.3. Plasmids. (Source: Di Pilato V, Antonelli A, Giani T, Henrici De Angelis L, Rossolini GM and Pollini S (2019) Identification of a Novel Plasmid Lineage Associated With the Dissemination of Metallo-β-Lactamase Genes Among Pseudomonads. Front. Microbiol. 10:1504. doi: 10.3389/fmicb.2019.01504, Creative Commons License).

When scientists first began studying bacteria and phages, they were naturally drawn to their genetic systems. It is evident that bacteria and phages must have genetic mechanisms as they display a steady appearance and function from one generation to the next. Bacteria reproduce asexually by cell division and expansion, in which one cell divides into two, much like unicellular eukaryotic species. This asexual reproduction is relatively easy to demonstrate experimentally. We can see that phages and bacteria both contain a variety of genetic mechanisms. These processes are fascinating due to the fundamental biology of these forms and serve as models, and sources of knowledge about the genetic mechanisms at work in all species. These forms are appealing to geneticists because they can be produced in enormous quantities due to their tiny size. This allows for the discovery and study of exceedingly rare genomic events that are difficult or impossible to study in eukaryotes. During sexual reproduction, two DNA molecules from two different sources are joined. Bacteria and eukaryotes differ significantly in that two complete chromosomes are rarely brought together in bacteria; instead, the union often consists of one complete chromosome plus a fragment of another. The first form of gene exchange to be investigated is conjugation, which is the interaction and fusion of two different bacterial cells. After fusion, a cell known as a donor occasionally gives genomic DNA to the recipient cell. It is possible that the full bacterial genome or only a small fraction of it was transmitted. Plasmids, which are autonomous extragenomic DNA elements, may be transmitted under specific conditions. These plasmids can transfer genomic DNA to the intended cell. No matter how it was introduced, any genetic segment inserted has the potential to recombine with the recipient's chromosome. By absorbing some DNA from the outside environment and incorporating it into its own genome, a bacterial cell can likewise change itself.

4.11 PHAGES

Some phages can also transfer DNA from one bacterial cell to another, where it can be incorporated into the chromosome, a process known as transduction. The transfer of DNA on a plasmid through transformation or transduction is known as horizontal transmission, a type of gene transmission that does not need cell division (Lerminiaux & Cameron, 2019). This term distinguishes this type of DNA transfer from vertical transmission, which is the transfer of DNA across subsequent bacterial generations. By contact, horizontal transmission can quickly transfer DNA through a bacterial community, much like how a disease does. Bacteria can quickly adapt to changing environmental circumstances due to horizontal transmission. The phages themselves may undergo recombination when two phages of various genotypes infect the same bacterial cell. Bacteria make ideal genetic model organisms because they proliferate quickly and take up little space. They can be grown either on a solid surface like an agar gel or in a liquid medium as long as the appropriate nutrients are supplied. Each bacterial cell divides asexually in multiples of 1, 2, 4, 8 cells, and so on, until the nutrients run out or there is too much toxic waste to support further growth. A small amount of a liquid culture can be pipetted onto a Petri plate containing solid agar medium and spread evenly with a sterile spreader during the process known as plating. Although the cells divide, they remain in a group because they can't move around much on the gel's surface. Once this mass contains more than 10^7 cells, a colony can be seen without the help of a microscope. There are numerous colonies on the plate that originated from different initial cells. Colony members known as cell clones have a single genetic progenitor in common. Mutant microbes are quite easy to cultivate. A good example is nutritional mutations. Wild-type bacteria are prototrophic because they can multiply and thrive on a substrate that is only composed of inorganic salts, a carbon source for energy, and water. Prototrophic cultures can be used to create auxotrophic mutants, or cells that cannot grow unless the media contains one or more cellular building blocks like adenine, threonine, or biotin. Another way that helpful mutants differ from their wild-type counterparts is in their ability to use a certain energy source. For example, the mutant (lac) cannot consume lactose and grow, whereas the wild-type (lac+) can. An inhibitor, such as the antibiotic streptomycin, does not stop resistant mutants from dividing and forming colonies; however, wild types are still vulnerable to the inhibitor. Geneticists can distinguish between different strains separately due to all of these mutants, which also offer genetic markers (marker alleles) for tracking genomes and cells in experiments.

By accident, Hayes discovered a subtype of the initial donor strain that, when bred with the recipient strain, did not produce recombinants. It seems that after losing its ability to transfer genetic material, the donor-type strain had changed into a recipient-type strain. Hayes discovered that this sterile donor variety may regain its ability to operate as a donor by interacting with other donor strains. In actuality, the donor capacity was swiftly and effectively transferred between strains during conjugation. It looked like some component had undergone an infectious transmission. He claimed that a genetic status is imposed on donor ability by a reproductive element (F).

Strains that are F+ are those that can donate and carry F plasmids. F-deficient strains are recipients rather than donors and are referred to as such. We now know a great deal more about F. It serves as an example of a plasmid, a tiny, circular DNA molecule that may grow in the cytoplasm on its own without the assistance of the host chromosome. The F DNA in the donor cell can produce a single-stranded copy of itself due to the unusual process known as rolling circle replication. A single-strand is reeled out as the circular plasmid moves. A pore allows this one strand to enter the recipient cell, where it joins with another strand to form a double helix. A copy of F thus stays in the donor while a new copy forms in the recipient. Most bacterial genomes are circular, in contrast to the eukaryotic nuclear chromosomes.

The seventeenth-century scientist Antony van Leeuwenhoek, who was perhaps the first to see bacteria and recognize their minute size, once stated that there are more bacteria living in a man's mouth than there are men in the entire kingdom. However, bacteriology didn't fully take off until the nineteenth century. Bacteriology was thrust into the emerging field of genetics in the 1940s when Joshua Lederberg and Edward Tatum discovered that a particular bacteria had a type of sexual cycle that included a crossing-over-like process. Escherichia coli was the organism, which was called after the German bacteriologist Theodore Escherich, who discovered it in the nineteenth century (Farré-Maduell & Casals-Pascual, 2019).

4.12 *E. COLI*-THE MODEL

The selection of *E. coli* was fortunate since it has shown to have various qualities that make it ideal for genetic research, not the least of which is that it is readily available because it lives in the gastrointestinal tract of humans and other animals. It lives in the stomach as a benign symbiont, but on rare occasions, it can also lead to diarrhea and urinary tract infections. The only chromosome in *E. coli* is 4.6 Mb. The functions of about 35% of the 4000 intron-free genes are unknown. The sexual cycle is made possible by the extragenomic plasmid F. Other plasmids contain genes, such as those that provide drug resistance, whose functions prime the cell for survival under specific circumstances. These plasmids have been altered to serve as gene vectors, also known as gene carriers, which are the fundamental units of the gene transfers at the center of modern genetic engineering. Since *E. coli* only has one cell, it reproduces by cell division. Due to its small size, *E. coli* may be grown in large quantities and is amenable to strict selection and screening for unusual genetic events. We must first grasp a few distinctive features of the recombination process in bacteria. Remember that recombination between two whole genomes only occurs in eukaryotes. On the other hand, it happens when two incomplete genomes, the exogenote and the endogenote derived from the Hfr donor and recipient cells, respectively. The cell now has two copies of the same DNA sequence, one of which is an exogenote component and the other of which is an endogenote component. As a result, the cell is currently a merozygote, a partly diploid. Bacterial genetics is merozygote genetics. A single crossover in a merozygote would break the ring, and prevent the generation of viable recombinants. To keep the shape, the number of crossovers must be precisely even. Both a broken circular chromosome and one with an even number of crossings are produced. Although these recombination events are sometimes referred to as double crossovers, the real chemical mechanism is somewhat different and more like an exogenote internal section invading an endogenote. Most of the time, additional cell growth gets rid of the fragment, which is the other byproduct of the double crossing. Recombination results in only one viable product from its reciprocal products. Since reciprocal exchange is another special feature of bacterial recombination, we must often omit its byproducts.

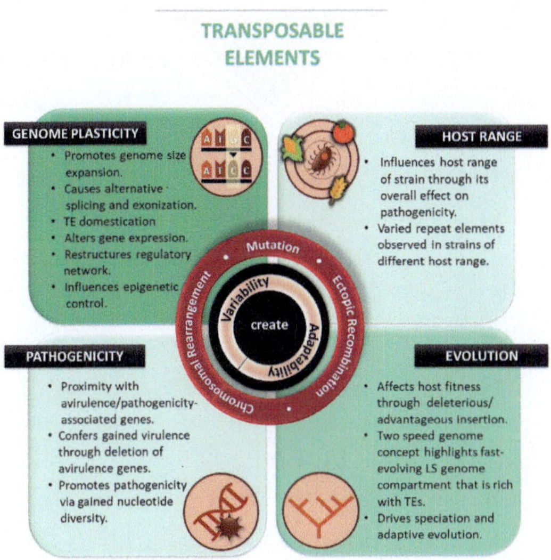

Figure 4.4. Transposons and their importance. (Source: Mat Razali N, Cheah BH, Nadarajah K. Transposable Elements Adaptive Role in Genome Plasticity, Pathogenicity and Evolution in Fungal Phytopathogens. *International Journal of Molecular Sciences*. 2019; 20(14):3597. https://doi.org/10.3390/ijms20143597, Creative Commons License).

4.13 TRANSPOSONS

Rarely, as a transposon (Figure 4.4) moves to a new location in the genome, other kinds of genes, such as alleles for drug resistance, can be enclosed between the ends of the transposon (Durfee et al., 2008)taking advantage of properties such as high DNA transformation efficiency and maintenance of large plasmids. The strain was constructed by serial genetic recombination steps, but the underlying sequence changes remained unverified. We report the complete genomic sequence of DH10B by using reads accumulated from the bovine sequencing project at Baylor College of Medicine and assembled with DNAStar's SeqMan genome assembler. The DH10B genome is largely colinear with that of the wild-type K-12 strain MG1655, although it is substantially more complex than previously appreciated, allowing DH10B biology to be further explored. The 226 mutated genes in DH10B relative to MG1655 are mostly attributable to the extensive genetic manipulations the strain has undergone. However, we demonstrate that DH10B has a 13.5-fold higher mutation rate than MG1655, resulting from a dramatic increase in insertion sequence (IS. Then, these genes are carried as cargo to their new places. Their attachment of a drug-resistance allele to a plasmid can occasionally result in the production of a R plasmid. Like F plasmids, many R plasmids are conjugative, successfully transmitting to a recipient cell during conjugation. R alleles can be transposed from conjugative plasmids to non-conjugative R plasmids, which never leave their host cells. As a result, plasmids enable the rapid spread of antibiotic-resistant genes among a population of bacteria. Although the spread of R plasmids aids in bacterial survival, it creates a problem for drug treatment since, as already mentioned, bacterial populations quickly acquire resistance to any new antibiotic created and given to patients.

Some bacteria can swap genes with one another by ingesting DNA fragments from the surrounding environment. Species-specific cells or cells from various species can both contribute DNA. The DNA has occasionally been secreted from living bacteria, and occasionally it has come from dead bacteria. The recipient's chromosome accepts the DNA and incorporates it. If this DNA has a different genotype than that of the recipient, the process known as transformation can permanently change the recipient's genotype. In the year 1928, Frederick Griffith discovered transformation in the *Streptococcus pneumoniae* bacteria. The transforming DNA is inserted into the bacterial chromosome by a method analogous to the double-recombination processes observed in Hfr-F crossings. Be aware that conjugation includes the transfer of DNA from one living cell to another by close contact as opposed to transformation, which requires discrete exogenous DNA fragments being absorbed by a cell through the cell wall and plasma membrane.

4.14 TRANSFORMATION

Transformation has shown to be a useful technique in a variety of bacterial research disciplines since it allows for the intentional alteration of a strain's genotype (Ndagi, A. Falaki, Abdullahi, M. Lawal, & E. Soliman, 2020). For instance, transformation is commonly used in genetic engineering. It has been shown that bacterial and eukaryotic cells can both be modified by methods that are similar. Transformation can be used to determine how closely two genes are related on a bacterial chromosome. Some DNA fragmentation is inevitable during transformation studies. Two donor genes that are close to one another on the chromosome have a good chance of occasionally being carried on the same piece of transforming DNA. Therefore, both will be applied, resulting in a dual transformation. A genome may take up both sections or fragments separately, producing a two-fold transformant, although this is unlikely. The frequency of double transformants will consequently equal the total frequency of single transformants. In other words, if genes are related, the fraction of double transformants will be higher than the total frequency of single transformants. The intricacy of the issue is unfortunately increased by a number of factors, the most important of which is that not all bacterial cells in a population can be altered. The viruses kill bacteria by infecting them. In groundbreaking research on the genetics of bacteriophages done in the middle of the 20th century, recent studies on tumor-causing viruses and other types of animal and plant viruses have been constructed. Thus, bacterial viruses have become a crucial model system. These viruses can be used in genetic analysis techniques. Two different phage genotypes can be crossed first to measure recombination and then map the viral genome. Some phages can obtain bacterial genes and transmit them from one bacterial cell to another through a process called transduction. Thus, transduction is now a method of genetic material transfer between bacteria in addition to the transmission of Hfr chromosomes, F′ plasmids, and transformation.

Joshua Lederberg and Norton Zinder tested for recombination in the *Salmonella typhimurium* bacterium in 1951 using the techniques that had been fruitful with *E. coli*. The two strains that the researchers used were phe, Trp, Tyr and Met, His. When both strains were plated on a basic media, none of them developed any wild-type cells. The proportion of wild-type prototrophs was only present in roughly 1 in 10^5 instances when the two strains were mixed. However, in this case, the researchers also found recombinants from an experiment in a U-tube where conjugation was prevented by a filter separating the two arms. They put out the hypothesis that something was passing genes from one bacterium to another. By changing the size of the filter's pores, they were able to determine that the agent responsible for gene transfer was the same size as the well-known Salmonella phage P22. The filterable agent and P22 also had the same antiserum sensitivity and immunity to hydrolytic enzymes. Lederberg and Zinder have thereby discovered a brand-new type of virus-mediated gene transfer. To characterize this procedure, they coined the term transduction. Rarely, virus particles will take bacterial genes and transfer them to the next host they infect throughout the lytic cycle. Since then, it has become clear that many microorganisms can transmit signals. To understand the transduction process, we must distinguish between the two different phage cycles. A virulent phage instantly lyses and kills the host. Temperate phages can remain in the host cell for a very long time without killing it. To replicate, either their DNA replicates independently in the cytoplasm, similar to a plasmid, or it fuses with the host chromosome. Phages that have been incorporated into the bacterial genome are known as prophages. Lysogenic bacteria are also referred to as lysogens because they harbor dormant phages. Rarely, in a lysogenic bacterium would the dormant phage activate, grow, and cause the host cell to spontaneously lyse. A resident temperate phage confers resistance to infection by other phages of that type. There are two types of transduction: generalized and specialized. Generalized transducing phages can transfer any section of the bacterial chromosome, in contrast to specialized transducing phages.

In 1965, H. Ikeda and J. Tomizawa provided some insight using the E. coli phage P1 (Sternberg & Maurer, 1991). They found that when P1 lyses a donor cell, the bacterial chromosome is cleaved. Rarely, newly generated phage particles mistakenly incorporate a piece of bacterial DNA rather than phage DNA into the phage head. This event gave rise to the transducing phage. A phage that has bacterial DNA in it has the potential to infect another cell. The bacterial DNA can subsequently be inserted into the receiving cell's chromosome via recombination. Since genes on any of the host genome's fragments can be transduced, this form of transduction must be of the generalized type. A phage family that demonstrates generalized transduction includes P1 and P22. P22 DNA inserts into the host chromosome, while P1 DNA is free and functions like a large plasmid. Generalized transduction can be used to acquire information on bacterial linkage when genes are close enough to one another for the phage to take them up and transduce them in a single piece of DNA. Think about the instance where we wanted to figure out how far met and arg were from one another in E. coli. After cultivating them on a donor met+ arg+ strain, we might allow P1 phages from this strain to be lysed and infect a met+ arg+ strain. Let's start by selecting one donor allele, say met+. Then it is estimated what percentage of arg+ colonies are also met+. Bacterial chromosomal maps have been created by combining the techniques of interrupted mating, recombination mapping, transformation, and transduction. New genetic markers are now routinely mapped into a segment of 10 to 15 map minutes using intermittent mating. Then, additional, closely related markers can be mapped in a more in-depth analysis employing P1 co-transduction or recombination.

DNA from a transposon carrying a resistance allele or another kind of selectable marker is transformed and then transferred into bacterial receivers without active transposons (Jackson, Reinaldo Camacho, Gicquel, & Guilhot, 2001). Transposons effectively insert at random, and those that land in the middle of a gene cause a mutation. A subset of all mutants with traits corresponding to the bacterial process will be the focus of the investigation. Transposons have the advantage of making it easier to identify and sequence mutant genes because their sequence is well-known. DNA replication primers are made using a template based on the known transposon sequence. The short sequence obtained can then be compared to the complete genome sequence using a computer. The results of this investigation reveal the gene's location and entire sequence. It is possible that a gene's function in a different organism has previously been established. It is therefore obvious that this approach provides a way of relating mutant phenotype to map position and potential function. Developments in bacterial and phage genetics have paved the way for molecular biology and cloning over the past 50 years (Figure 4.5). Early on, it was observed that several bacterial strains exchanged genes and recombined. However, in bacteria, genetic material can only be transferred from a donor cell (F+ or Hfr) to a recipient cell in one direction, as in the instance of *Escherichia coli*. The donor's capability is determined by the presence of a certain kind of plasmid termed a fertility factor (F) in the cell. Sometimes the free F factor present in F+ cells integrates with the *E. coli* genome to form a Hfr cell. A donor chromosomal fragment may then enter a recipient cell and later reconnect with the recipient chromosome when this occurs. Because the F factor can insert at many sites on the host chromosome, early researchers were able to piece together the transferred portions to show that the *E. coli* chromosome has a single circular, or ring, structure.

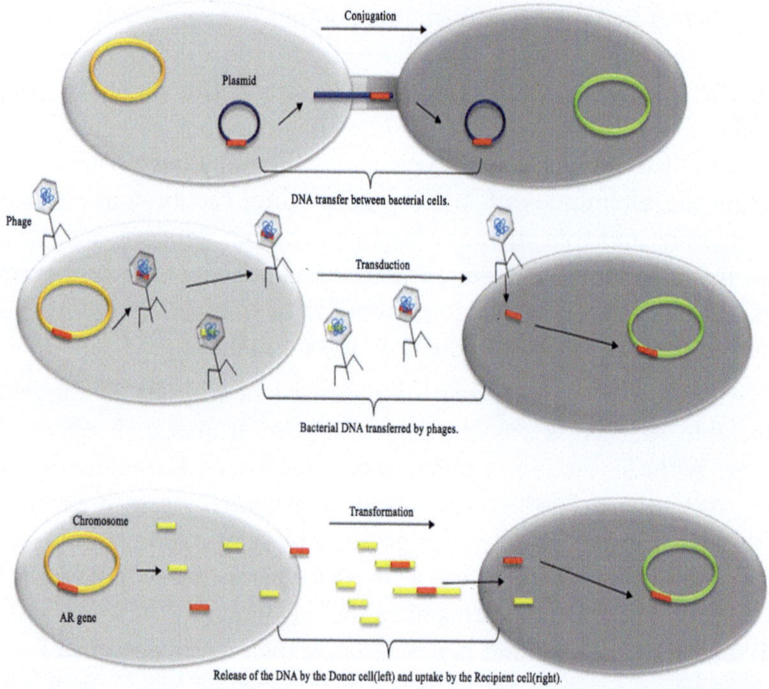

Figure 4.5. Transmission of traits among microbes. (Source: Vrancianu CO, Popa LI, Bleotu C and Chifiriuc MC (2020) Targeting Plasmids to Limit Acquisition and Transmission of Antimicrobial Resistance. Front. Microbiol. 11:761. doi: 10.3389/fmicb.2020.00761, Creative Commons License).

4.15 PLASMIDS

R plasmids usually house their antibiotic-resistance alleles inside of a transposon, a mobile element. Extensive drug resistance to essential medical therapies is caused by the rapid spread of plasmids. Such organic plasmid variants have evolved into important cloning vectors, useful for identifying and studying genes in all organisms. DNA strands that enter bacterial cells through the extracellular space can transfer genetic characteristics from one bacterial cell to another. The process of transformation in bacterial cells provided the first evidence that DNA is the genetic material. For transformation to occur, DNA must be absorbed into the recipient cell, and this requires recombination between the recipient chromosome and the incorporated DNA. Viruses that may infect bacteria are known as bacteriophages. By entering bacterial cells and using their metabolic activities to make progeny phages that burst the host bacterium, chromosomes are one way for phages to infect bacteria. The new phages can then infect more cells after that. When two phages of different genotypes infect the same host, recombination between their chromosomes might take place. Lysogeny, a different kind of infection, includes the injected phage lying latent inside the bacterial cell. The host chromosome frequently fuses with this dormant phage (prophage), which then replicates alongside it. Either spontaneously or in reaction to the correct conditions, the prophage can get activated from its latent state and lyse the bacterial host cell. A phage is able to transmit bacterial genes from a donor to a recipient.

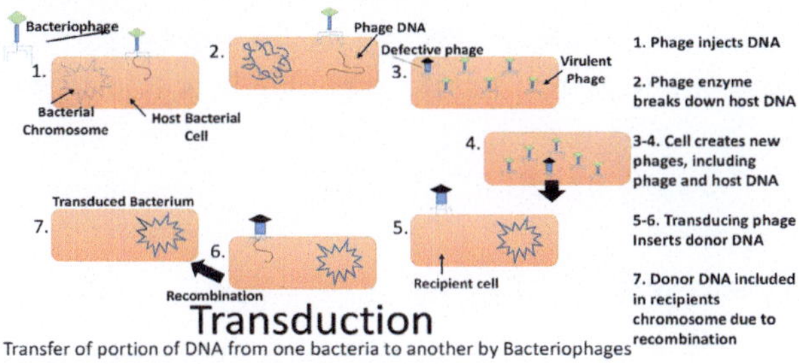

Figure 4.6. Generalized transduction. (Source: https://en.wikipedia.org/wiki/File:Transduction_image.pdf, Creative Commons License).

4.16 TRANSDUCTION

Generalized transduction (Figure 4.6) occurs after lysis and involves random host DNA integrating by itself into the phage head (Birge, 2000). During specialized transduction, if the prophage is erroneously excised from a specific chromosomal locus, the resulting phage head contains both specific host genes and phage DNA. A complete genome sequence acts as a physical map for many bacterial species. This physical genome map can be used to identify a mutation's precise location. To produce the right mutations, transposons must first be inserted (insertional mutagenesis). The DNA sequence surrounding the inserted transposon is then matched to a sequence from the physical map. This method provides the region, sequence, and maybe even function of the desired gene. The researcher starts with a collection of mutants and crosses each mutant with the wild type to see if a mutant shows single-gene inheritance. A collection of genes that are all involved in the development of the researched trait would be identified as a result of the information gathered from such a study. The researcher might in certain cases be able to determine precise biochemical roles for many of the genes by comparing the gene sequences of other animals. The next and more challenging step is figuring out how the genes in a set interact to determine phenotype. With the use of genome chips, this type of research is now possible.

The set of RNA transcripts present during a certain developmental phase can serve as a marker for the genes involved. Gene interactions and their significance in influencing phenotype can be inferred via genetic analysis. Only a small fraction of the tens of thousands of ways to alter the sequence of a gene will result in a mutant allele, which can appear in any population. The known mutant alleles of a gene and its wild-type allele are referred to as multiple alleles or an allelic series. One of the typical tests performed on a new mutant allele is to identify whether it is dominant or recessive. Understanding the fundamentals of dominance and recessiveness is helpful in coping with the new mutation and can offer insight into how the gene functions.

A fully dominant allele will be visible in the phenotype when only one copy is present, as in a heterozygote, but the alternative allele will be completely recessive. A/A Equals A/a at the phenotypic level because the homozygous dominant cannot be distinguished from the heterozygote in full dominance. As was previously mentioned, several single-gene human diseases, including phenylketonuria (PKU), have dominant wild-type alleles. While the wild-type allele is fully dominant in some single-gene disorders, such as achondroplasia, it is recessive in other single-gene disorders. The disease PKU is a good example of a recessive mutation. PKU is caused by a defective allele in the gene that codes for the phenylalanine hydroxylase enzyme (PAH).

In the absence of typical PAH, phenylalanine that is consumed through food is not broken down and accumulates in the body. Under these conditions, phenylalanine is converted into phenylpyruvic acid, which enters the bloodstream and travels to the brain where it inhibits normal brain growth and causes mental retardation. Many molecular pathways act as mediators of dominance. Another important class of dominant mutations are dominant negative mutations. A homodimeric protein unit, which is a protein made up of two identical units, can occasionally be the product of a gene. One example of a mutation that might act as a dominant negative is the gene that codes for the collagen protein. Certain mutations in this gene have been linked to osteogenesis imperfecta in humans (brittle-bone disease). Three intertwined monomers make up the connective tissue protein collagen (a trimer). In the mutant heterozygotes, the faulty protein wraps around one or two healthy proteins, causing the trimer to deform and malfunction.

Chapter

5

HUMAN GENETICS

CONTENTS

5.1 INTRODUCTION

Uncontrollable movements, behavioral abnormalities, and changes in thinking are all symptoms of Huntington's disease (HD) (Roos, 2010) behavioral and psychiatric disturbances and dementia. Prevalence in the Caucasian population is estimated at 1/10,000-1/20,000. Mean age at onset of symptoms is 30-50 years. In some cases symptoms start before the age of 20 years with behavior disturbances and learning difficulties at school (Juvenile Huntington's disease; JHD. Because it is dominant, each affected person's offspring only has to inherit one defective copy of the gene in order to contract the disease. Typically, symptoms appear in early adulthood, and 15 to 20 years later, the patient dies. Only 10% of HD patients are under the age of 20. Single-gene disorders like cystic fibrosis (CF) and Huntington's disease (HD) (Figure 5.1) affect families in ways known as modes of inheritance.

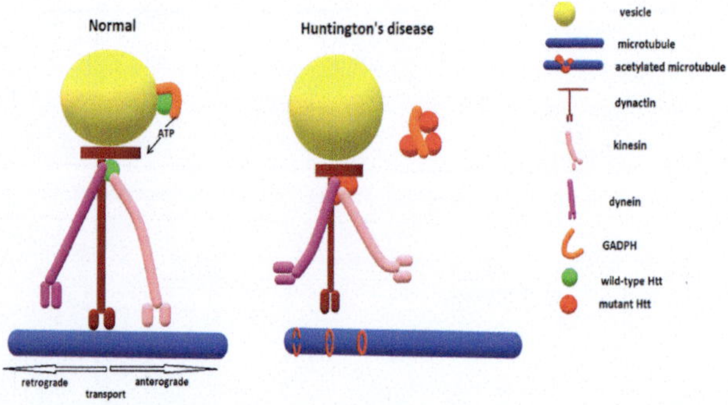

Figure 5.1. Huntington's disease. (Source: Jurcau A. Molecular Pathophysiological Mechanisms in Huntington's Disease. *Biomedicines*. 2022; 10(6):1432. https://doi.org/10.3390/biomedicines10061432, Creative Commons License).

Knowing these patterns enables one to anticipate the likelihood that a mutation will be inherited by certain relatives. Given that HD is autosomal dominant, it affects both sexes and occurs every generation. Contrarily, CF is autosomal recessive, which implies that both sexes are affected and that carriers who do not exhibit symptoms might cause the disease to skip generations. Today's genetic tests, such as exome and genome sequencing, can identify the single-gene diseases we carry. In recognition of the scientist who discovered the two rules of inheritance that govern how these qualities are passed down from one generation to the next, single-gene features are also known as Mendelian traits. Mendel's laws are still relevant in the era of sequencing since understanding family links frequently affects who to sequence, why to sequence them, and how to interpret the findings. The first person to investigate the fundamental logical principles that allow for the prediction of inheritance of specific features was Gregor Mendel. He recognized patterns in the natural variation of plants. Mendel studied the units of heredity that convey traits from one generation to the next by growing pea plants.. Mendel's rules of heredity have not only stood the test of time, but also explain how traits are transmitted in any diploid species, despite the fact that he had no knowledge of DNA, chromosomes, or cells.

5.2 PHENOTYPE AND GENES

Mendel crossed and recorded features in 24,034 plants over multiple generations between 1857 and 1863. He came to the conclusion that the plants transmitted unique units based on the regular ratios of attributes in the progeny. He came up with two theories to describe how inherited qualities are passed down. Few people read Mendel's discoveries when they were published in 1866. Peas are excellent for examining heredity since they are simple to cultivate, mature quickly, and have a variety of features that may be divided into distinct categories. The first generation is referred to as the parental generation, or P1, the second generation is referred to as the F1 or first filial generation, the third generation is referred to as the second filial generation, or F2, and so on. Mendel's initial studies focused on features with two expressions, such as short and tall, for example. In order to ensure that the plants always bred true, or produced the same expression of the trait, he started with cultivars that had been grown over many generations. In order to create artificial pollination, he set up every possible combination of tall and tall, short and short, and tall and short plants. Hybrids, or offspring that inherit a different gene variant (or allele) from each parent, are produced when plants with one phenotypic variant cross with plants with the alternate. Mendel observed that short plants that were crossed with other short plants were true-breeding, consistently generating short plants, which is the intended trait. While some tall plants were true breeders, others that were crossed produced short plants in around a quarter of the following generation. Tallness seems to mask shortness in some plants. A dominant trait suppresses a recessive trait, and vice versa.

Children born to an obese male and woman are also obese. The same issue affects a man who has an alcoholic father. Even single-gene diseases can be altered by the environment, other genes, or both. For instance, a child with cystic fibrosis inherits a single gene problem, but their experiences depend on the gene variants they have, additional genes that affect their immune systems, the infections they are exposed to, and the quality of the air they breathe. Either a trait is single-gene (Mendelian or monogenic), or it is polygenic. A polygenic trait, as its name suggests, represents the actions of several genes. Single-gene and polygenic traits can both be multifactorial, which indicates that environmental factors play a role in their development. Lung cancer has multiple risk factors. Purely polygenic traits, or those that are entirely unaffected by the environment, are extremely uncommon. Common qualities including height, skin tone, body weight, numerous diseases, and behavioral disorders and dispositions are all examples of polygenic multifactorial traits. The functioning of the brain, rather than another organ, is what distinguishes behavioral traits from other sorts of qualities. For multifactorial traits, genes are no more complicated than those of other traits. They adhere to Mendel's principles but only partially contribute to a trait, making them harder to track. Smoking-related lung cancer serves as an example of how complicated multifactorial features can be. If a person never smokes and just breathes fresh air, certain genetic variations that raise the chance of getting cancer and nicotine addiction may not ever occur. Several genes have cumulative effects in a polygenic multifactorial disorder. Although each gene imparts a certain level of sensitivity, the contribution of these genes is not always equal. An uncommon allele may have a significant impact, whereas multiple common alleles may each have a minor effect. For instance, three genes considerably increase the chance of type 2 diabetes mellitus, but other genes have less of an impact. A phenotype that was previously believed to be caused by the actions of a single gene may instead be caused by several genes. For migraine, this is the situation. The sensitivity to sound is influenced by a gene on chromosome 1, the pulsing headache and light sensitivity are caused by a gene on chromosome 5, and nausea and vomiting are influenced by a gene on chromosome 8 (Nyholt et al., 2005)complex disorder that shows strong familial aggregation. Using latent-class analysis (LCA.

5.3 ENVIRONMENTAL FACTORS

In addition, some people's migraines are triggered by environmental factors like consuming particular foods. For the heart and blood vessels (the cardiovascular system) to pump blood, numerous cell types and functions must work together, and many genes support the system. Even for cardiovascular disorders caused by a single gene, the environment has substantial effects. For instance, the severity of single-gene clotting diseases is influenced by vitamin K consumption, which is required for blood to clot. One in three people have cardiovascular disease. The transfer of lipids, blood clotting, blood pressure, and how well white blood cells adhere to blood vessel walls are just a few of the ways that genes influence the heart and blood vessels. Only until lipids are joined to proteins to form big molecules known as lipoproteins can they circulate in the blood. Apolipoproteins, the protein components of lipoproteins, are encoded by a number of genes. Others transport lipids to the liver, where they are converted into biochemicals that the body can excrete more readily. Some types of lipoproteins transport lipids in the blood to tissues, where they are needed. Smokers have a threefold increased chance of having a heart attack due to the apolipoprotein E gene (the E4 allele). A balance must be struck between having enough lipids inside cells and having not too many outside cells in order to maintain a healthy heart and blood vessels. Many genes that define lipid-processing enzymes, lipid-transporting proteins, or receptor proteins that let lipids into cells regulate the lipid levels in the blood and tissues. The walls of the smallest blood vessels are lined with an enzyme called lipoprotein lipase, which breaks down fat packages expelled from the small intestine and liver. High-density lipoproteins (HDLs) trigger lipoprotein lipase, which degrades low-density lipoproteins (LDLs). Low LDL is used as a diagnostic of heart health since it is linked to a healthy cardiovascular system in many families. Before cardiologists realized what geneticists had long known that some families without cardiac problems have low HDL, typically a warning sign, high HDL was once frequently regarded as a diagnostic for health. Statin medications, which block the enzyme needed by the liver to produce cholesterol, are available to patients with increased LDLs that diet and exercise are unable to control (Charlton-Menys & Durrington, 2008) receptors and transporters in cholesterol biosynthesis and transfer and the development of drugs for the regulation of cholesterol metabolism. The discovery of the low-density lipoprotein (LDL.

5.4 INTERESTING FACTS

Health also depends on the blood's fluidity. Usually in blood veins in the heart or legs, excessively active clotting factors or sticky white blood cells can cause clots to develop that obstruct blood flow. Dangerous bleeding results from poor clotting. Clotting (Figure 5.2) is genetically governed as clotting factors are proteins. Familial cholesterolemia is an example of how cardiovascular disease has several underlying causes. Heart attacks usually occur in young adults who inherit one mutant allele of the LDL receptor gene, and those extremely rare people who receive two mutations die much sooner.

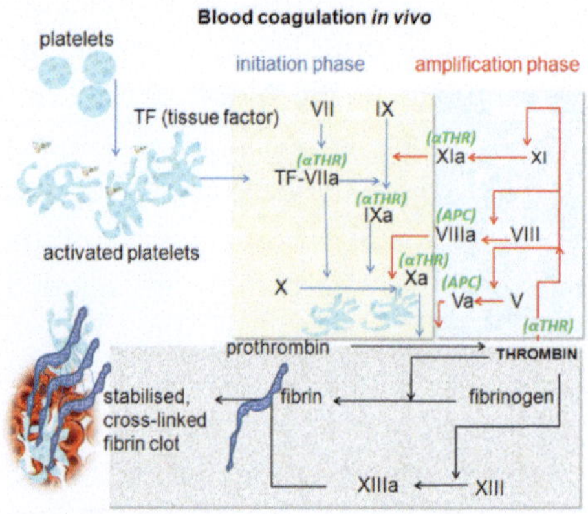

Figure 5.2. Clotting. (Source: By Dr Graham Beards - Own work (Based on Figure 7.5 on page 167 of "Blood Science: principles and pathology"by Andrew Blann and Nessar Ahmed, Publishers Wiley Blackwell (2014) ISBN:978-1-118-35146-8), CC BY-SA 3.0, https://commons.wikimedia.org/w/index.php?curid=19772662, Creative Commons License).

The genome has genes that affect cholesterol deposition in at least twelve other locations, in addition to food. Through diet and exercise, many people with dangerous genes can prevent symptoms. Many gene alleles are found by genetic tests, and each one increases the risk of having cardiovascular disease. More than 95 genes contribute to inherited variation in blood cholesterol and triglyceride levels, while more than 50 genes influence blood pressure. Gene expression tests can reveal which cholesterol-lowering medications are most likely to be efficient and tolerated for a certain patient. Multigene test analysis using computers compensates for environmental factors that may be changed, like exercising, quitting smoking, and keeping a healthy weight. For a polygenic trait, the interaction of many genes frequently results in a phenotype, also known as a quantitative trait. Quantitative trait loci, or QTLs, are DNA sequences that influence polygenic traits. If a trait is both multifactorial and polygenic, it is continuously changing. In other words, several genes contribute to the phenotype's ongoing variance. Mendel's rules are followed by the individual genes that confer polygenic traits, but when they are combined, they do not result in single-gene phenotypic ratios.

Although none of them are dominant or recessive to one another, they all contribute to the phenotypic. An "all-or-none" phenotype, such as "normal" versus "affected," is commonly provided by single-gene features instead of discrete or qualitative qualities. As our many variations in hair color, body weight, and cholesterol levels show, polygenic traits vary in groups. A polygenic characteristic is influenced more by some genes than others. Depending on the specific way an allele modifies a protein encoded in a gene and how prevalent it is in a population, alleles within genes can have varying effects. For instance, a mutation in the gene for the LDL receptor significantly boosts the level of blood serum cholesterol. However, because this mutation affects less than 1% of people in the majority of populations, it has a very small impact on the diversity in cholesterol levels at the population level. The mutation, however, has a significant effect on the individual who possesses it.

Although polygenic expression is continuous, we can classify people into groups and estimate the frequencies of the groups. This produces a bell-shaped curve when we plot the frequency for each phenotypic class. Dermal ridges are folded patterns of elevated skin on the fingertips that line up to form loops, whorls, and arches. This design resembles a fingerprint. In order to recognize and distinguish people, a process known as dermatoglyphics compares the number of ridges that make up these patterns. Due to the peculiar ridge patterns present in some diseases (Down syndrome), dermatoglyphics is a component of genetics. Dermatoglyphics is also used in forensic fingerprint analysis (Hazarika & Russell, 2012). The amount of ridges in a fingerprint is mostly determined by genes, but the environment can also have an impact. The ridge pattern can change throughout weeks 6 through 13 of pregnancy as the fetus presses the finger and toe pads to the wall of the amniotic sac. It is due to this early environmental influence that some identical twins show differing fingerprints from one another while having the same genes. A measurement known as a total ridge count, which counts the number of ridges in whorls, loops, or arches, is used to quantify a fingerprint. Males typically have 145 total ridges, while females typically have 126. The bell curve of a constantly evolving attribute can be seen by plotting the total ridge count. People who do not eat enough do not grow as tall as they are genetically capable of. This is a clear example of how the environment affects height. Skin, hair, and iris pigmentation are all influenced by more than 100 genes. Various people have different levels of melanin pigmentation in their skin. Melanocytes, which are cells that carry melanin in packages called melanosomes, are found in both the skin and the eye. Melanocytes spread pigment granules throughout the skin layers by extending between the skin cells that resemble tiles. Some melanin leaves the melanocytes and moves into the higher layers of the skin's dense cells. Here, melanin fragments provide color when the skin cells are propelled upward toward the skin's surface and stem cells beneath them divide. UV radiation is shielded from DNA damage by the pigment. Melanin synthesis is accelerated by sunlight exposure. Despite being oversimplified, this example shows how multiple genes can influence a variable trait.

Per square centimeter of skin, every one of us has roughly the same number of melanocytes. Due to differences in melanosome size, number, and pigment distribution density, each person has a unique color of skin. The quantity and distribution of melanin particles in the cells of the uppermost skin layers determine the various skin colors. One physical characteristic that is used to differentiate races is skin color. Since skin color is just one of thousands of features whose frequencies fluctuate in various populations, the concept of race as defined primarily by one's skin color is more of a social construct than a biological one. When referring to nonhumans, races are groups within species that are characterized by various allele frequencies from a genetic perspective.

Compared to other mammals, humans really have much less variation in appearance. We may categorize people based on their skin tone because it is an easy method to differentiate them from one another, however, this characteristic is not a valid historical indication. When many genes are taken into account, the idea of race based on skin color is no longer valid. In other words, two persons with very dark complexion may have less in common with one another than they would with someone with very light skin. Sub-Saharan Africans and Australian aborigines, for instance, both have dark complexions, despite having quite different genetic traits. Their dark skins might be an adaptation to living in a warm, sunny climate rather than a result of recent common ancestry. 93 percent of different genetic features are distributed evenly among all people, regardless of skin color. According to DNA analysis, dividing people into ancestry and skin color categories makes more biological sense. In a practical sense, racial groups do have differing occurrences of certain diseases because of the tendency to choose spouses within a group, which keeps specific genes. In a genetic sense, the concept of race based on skin color has limited validity (Cooper, 1984). However, socio economic injustices like unequal access to a healthy diet or medical care may also contribute to racial inequalities in disease frequency. Race-based prescribing has been developed as a result of observations that certain diseases are more common in populations of particular races. For instance, because African Americans have a higher incidence of heart disease and hypertension than persons of other races do, these medications are marketed to them. On a personal level, however, if a treatment choice is made based on a characteristic unrelated to how the body reacts to a drug, a white person may be rejected a medication that would work while a black person may be offered one that does not.

5.5 DRUGS AND GENOTYPE

Instead of prescribing medication based on a patient's skin tone, it is more accurate to use the patient's genotypes, which determine whether or not a given medication will work or have adverse effects. For instance, in 354 individuals representing blacks (Bantu, Ethiopian, and Afro-Caribbean), whites (Norwegian, Armenian, and Ashkenazi Jews), and Asians, researchers compiled 23 markers for genes that regulate drug metabolism (Chinese and New Guinean). These response groups did not at all resemble the groups based on skin color. Instead, the genetic markers fell into four highly separate groups that predict which of various blood thinners, chemotherapies, and painkillers will be beneficial. Knowing the mode of inheritance such as autosomal dominant or recessive allows one to use Mendel's laws to forecast the likelihood that a single-gene characteristic would recur in a family. Geneticists utilize empiric risk, which is based on incidence in a particular group, to forecast the likelihood that a polygenic multifactorial trait will develop in a specific individual. The frequency of an event, such as the number of newly diagnosed instances of a disorder per year in a population of known size, is referred to as its incidence. Prevalence is the percentage of people in a population who are afflicted with a specific disorder at a given time, like throughout the course of a year. Empiric risk is a population statistic based on observation rather than a calculation. The population may be large, like an ethnic group or community, or it may be more genetically defined, like a group of families affected by cystic fibrosis. Empiric risk rises with the disorder's severity, the number of affected family members, and a person's degree of relationship with those who are afflicted.

For instance, the possibility that a kid will be born with a neural tube defect (NTD) is predicted using empiric risk (Godbole et al., 2011). The risk of bearing a fetus with an NTD is around 1 in 1,000 in the United States (0.1 percent). The risk is roughly 1 in 1,000 for those with English, Irish, or Scottish heritage. For any ethnic group, the risk of recurrence rises to 3% if a sibling has an NTD, and the risk to a third child is significantly higher if two siblings are affected. If a trait contains a genetic component, it makes sense that the more closely related two people are, the more likely it is that the second person will also have the trait because they share more genes. This reasoning is supported by empiric risk studies.

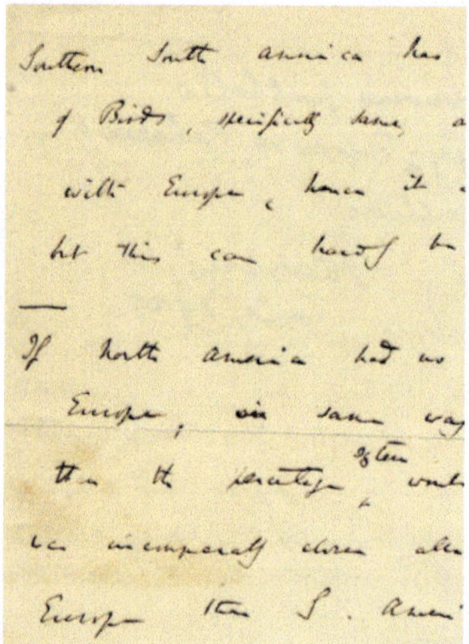

Figure 5.3. A handwritten manuscript note by Charles Darwin from *Natural Selection* portfolio (Cambridge University Library). (Source: Eldredge N. Darwin's *Other* Books: "Red" and "Transmutation" Notebooks, "Sketch," "Essay," and *Natural Selection*. *PLoS Biol.* 2005;3(11):e382. doi:10.1371/journal.pbio.0030382, Creative Commons License).

Empiric risk is useful for determining risks for diseases with poorly understood transmission patterns because it is completely dependent on observation. For instance, several multifactorial diseases are more common in one sex than the other. Males are five times more likely than females to have pyloric stenosis, an overgrowth of muscle at the junction between the stomach and small intestine.

Shortly after delivery, the issue needs to be surgically treated to prevent the child from being unable to digest food. According to empirical research, the probability of recurrence is 3.8 percent for the brother of an affected brother but 9.2 percent for the brother of an affected sister. Therefore, an empiric risk is one that is based on actual observations. It is not necessary to identify the disease's root cause.

5.6 DARWIN: A VITAL LINK

According to Charles Darwin, (Figure 5.3) some variation in a trait results from inborn differences between populations, and some from variations in external effects. Heritability, abbreviated H, quantifies how much the phenotypic fluctuation in a given population at a given time and may be attributed to genetic variations. Empiric risk and heritability differ in that heredity focuses on the genetic component of variation in a characteristic, whereas empiric risk may be caused by non-genetic sources. Heritability does not refer to the percentage of the trait that can be assigned to genes, but rather to the amount of variation in a trait that can be ascribed to genetics. For a variable whose variability is entirely the product of gene activity, such as in a population of laboratory mice living in the same environment, heritability is equal to 1.0. Without environmental variation, its expression in a population is determined only by genetic differences. However, variation in most features is caused by variations in genes and environmental factors. As the environment changes, heritability also changes. In the winter, for instance, when light exposure is less likely to enhance melanin formation, the heritability of skin color is higher. The same characteristic may be highly heritable in two populations, but due to long-term environmental variations, some variants may be far more prevalent in one community. African populations, for instance, have darker complexion than Scandinavians who don't get enough sunlight. To calculate heritability, scientists employ a variety of statistical techniques. One approach is to compare the actual proportion of pairs of individuals who are related in a certain way with the expected proportion of individuals who would share the trait if it were inherited in a Mendelian manner.

Due to the meiotic process, a parent and child share 50% of their DNA (Myers et al., 2011). Due to the probability of receiving each gene's allele from each parent, siblings typically share 50% of their genes. When calculating risks, genetic counselors use the terms primary (1°), secondary (2°), and tertiary (3°) relatives. The value of 1 in 2 or 50% between siblings and parent-child pairs can be used to trace and quantify the proportion of genes shared by people related to one another in various ways in lengthy or complex pedigrees. A set of 100 sibling pairs would be predicted to share approximately 50% of the same phenotype if the heritability of a trait is quite high, as siblings typically share 50% of their genes.

Heritability for height is a trait that reflects the impact of nutrition and the environment. For instance, 40 sibling pairs in a population of 100 could have the same amount of inches in height. This group of sibling pairs has an 80 percent heritability for height, which is calculated as the observed phenotypic variation divided by the anticipated phenotypic variation under a non-environmental scenario. The cumulative effects of various genes' recessive alleles account for the genetic variance for a polygenic trait. A few dominant alleles can have a significant impact on the phenotype for some traits, but because they are uncommon, they have little effect on the heritability of the trait. This holds true for cardiac disease brought on by a malfunctioning LDL receptor. If just mutations that modify DNA base sequences are examined, and not also copy number variants (CNVs, which are differences in the numbers of copies of a DNA sequence), heritabilities for some traits or diseases may be underestimated.

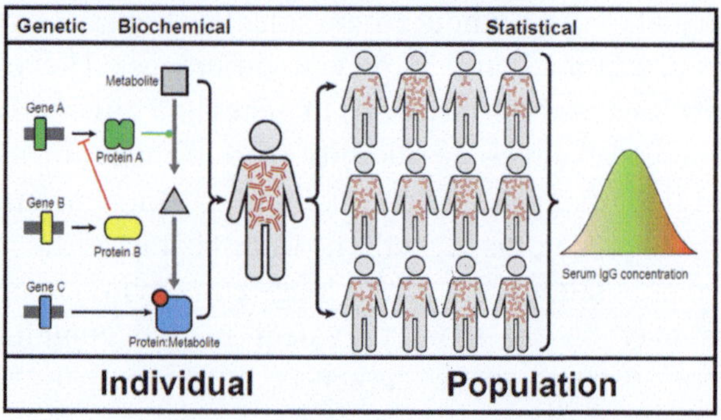

Figure 5.4. Epistasis. (Source: Ameratunga R, Woon S-T, Bryant VL, Steele R, Slade C, Leung EY and Lehnert K (2018) Clinical Implications of Digenic Inheritance and Epistasis in Primary Immunodeficiency Disorders. Front. Immunol. 8:1965. doi: 10.3389/fimmu.2017.01965, Creative Commons License).

5.7 EPISTASIS

Epistasis, (Figure 5.4) which is the interaction of alleles from various genes, can also affect heredity (Phillips, 2008)including the functional interaction between genes, the genetic outcome of mutations acting within the same genetic pathway, and the statistical deviation from additive gene action. Converging interests across genetics suggests that it is now time to develop a more unified view of epistasis and gene interactions.One of the traditional uses of epistasis analysis has been to order genes within developmental and metabolic pathways. These approaches have recently become much more systematic through the use of high-throughput genetic screens, especially in yeast.

These studies show that gene interactions are ubiquitous and can be used to help understand the structure of complex genetic networks.The major limitation of comprehensive analyses of gene interactions is the total number of interactions that must ultimately be tested, which grows at approximately the square of the number of genes (for example, >18 million interactions in the Saccharomyces cerevisiae genome. Geneticists compute a narrow heritability that only takes additive recessive effects and a wide heritability that additionally takes into consideration the effects of uncommon dominant alleles and epistasis to account for the fact that various genes affect a trait to varying degrees.

For instance, the narrow heritability of LDL cholesterol is 0.36, whereas the broad heritability is 0.96, demonstrating the significant influence of a rare dominant allele. Agriculture benefits from an understanding of multifactorial inheritance. Breeders need to understand if environmental or genetic factors play a role in the variability of features including birth weight, milk output, and egg hatchability. Knowing whether the genetic influences are additive or epistatic is also important. By modifying the conditions under which animals are kept and crops are cultivated and by arranging crosses between certain individuals, the breeder may regulate both the environment and genetic effects. It is challenging to study multifactorial features in humans since data must be gathered from many families. Adopted individuals and twins, however, can assist geneticists in unraveling the genetic and environmental components of the variability of multifactorial traits. In most cases, an adopted person and the adoptive family share contextual factors but few genetic variations. Adopted people, on the other hand, have DNA in common with their biological parents but not the same environment. Because of this, biologists believe that while similarities between biological parents primarily represent genetic factors, similarities between adopted people and their adoptive parents primarily reflect environmental influences. It is possible to determine how genetics and environment interact by looking up information on both sets of parents.

A database of all adopted children in Denmark and their families from 1924 to 1947 was used in several early adoption research. In one study, the causes of death among biological and adoptive parents as well as adopted children were compared. The likelihood of an adopted kid dying of infection at a young age was five times higher if a biological parent died of infection before age 50 than the likelihood of an adoptee dying of infection at that age in the general population. This may be due to the increased vulnerability to specific infections caused by inherited immune system gene variations.

This theory is supported by the finding that the likelihood of an adopted person dying young from infection did not correspond with the chance of the adoptive parents dying from infection before the age of 50. The majority of a person's lifespan is determined by genetics, but there is evidence that environmental factors also have a role. For instance, if adoptive parents died from cardiovascular disease before the age of 50, their adopted children had a threefold increased risk of dying from heart and blood vessel disease compared to the general population.

Studies that employ twins to distinguish between the genetic and environmental contributions to a trait are more insightful than those that use adopted people as their subjects. Twins have been used to examine the genetic influences on features since 1924, when German dermatologist Hermann Siemens discovered that identical twins and fraternal twins had considerably higher grades (Rende, Plomin, & Vandenberg, 1990). This observation led him to the hypothesis that genes play a role in intelligence. A characteristic is at least partially inherited if both members of identical (monozygotic or MZ) twin pairs experience it more frequently than both members of fraternal (dizygotic or DZ) twin pairs. Discordant twins are those who differ from one another in that attribute.

Researchers observe individuals' body language and facial expressions while also interviewing them about their interests, anxieties, and superstitions. Separated twins and triplets offer a natural experiment for separating nature from nurture. Particularly if their backgrounds have been very dissimilar, heredity can be blamed for many of their shared characteristics. Since their genes are identical (in the case of MZ twins and triplets) or comparable (in the case of DZ twins and triplets), their differences typically result from variances in education and growth. Even when they are raised in completely different adoptive families, some MZ twins who were split apart at birth and later reunited seem very similar. Eccentricities stand out in particular. Adoptive or biological siblings do not usually live in the same household. Each person's perception of contextual effects like parental affection and punishment is influenced by differences in their sex, general health, experiences with peers and at school, temperament, and personality.

5.8 GENOME-WIDE ASSOCIATION STUDIES

A genome-wide association study (GWAS), which compares extensive sets of genetic markers across the genome between two sizable groups of individuals, one with a certain trait or condition and the other without it, is a more recent method for analyzing multifactorial traits and disorders. Researchers can find genes that contribute to the phenotype by identifying regions of the genome that are significantly more prevalent in individuals who have the characteristic or condition. Genetic markers are used in genome-wide association studies. The DNA base sequence is described by single nucleotide polymorphisms (SNPs) and copy number variants (CNVs). An SNP is a location in the genome where at least 1% of a population contains a variant DNA base. A CNV is a DNA sequence that differs in the number of times it repeats in different people. A CNV is another means to identify individuals but does not impart information in the same way that a gene that codes for a protein does. Forensic applications can benefit greatly from CNVs (Zubakov et al., 2018)as it allows evaluating the crime relevance of such traces and enables reconstructing the sequence of crime events. Messenger RNA and micro RNA markers are useful for forensic tissue identification, but provide challenges for linking RNA-identified cell/tissue types with DNA-identified trace donors, especially in mixed traces. DNA methylation markers overcome this problem, but provide technical challenges due to the DNA treatment required by most analysis methods. Here we introduce a novel type of DNA markers for forensic tissue identification analysed without prior DNA treatment, namely copy number variation (CNV.

Genome-wide association studies also make advantage of gene expression patterns. A genome-wide association study needs at least 100,000 markers to be statistically significant. Informational markers are those that link to a characteristic or condition. Genome-wide association studies often make use of a million SNPs or more, divided into half a million or so haplotypes. A GWAS is an iterative focus on certain regions of the genome that are at least partially responsible for a trait. Typically, the DNA of a control group and a group of persons with the same disease or trait is extracted and genotyped for the 500,000 SNPs. The SNPs in the group that are specifically shared by the characteristic or condition are found by statistical techniques. The SNPs are strengthened by repeating the method on other populations. To be sure that an SNP relationship is related to the trait of interest and not another region of the genome that people in one population share because of their shared ancestry, it is crucial to confirm it in various population groups. Researchers monitor a sizable group of people in a cohort study throughout time and assess numerous facets of their health. Each participant in a case-control study is matched with a participant in the control group who shares as many characteristics as feasible, including age, sex, level of activity, and environmental exposures. The existence or absence of the disease or trait is then linked to SNP variations. Researchers search genomes for SNPs that are typically shared by siblings with the same ailment but are less frequently shared by siblings who do not have the same condition. Genes in these genomic areas might be involved in the disorder.

Homozygosity mapping is carried out in families where the parents are related. Children whose parents share ancestors have more homozygous areas in their genomes than other children, which increases the possibility that they have inherited two copies of a mutation that increases susceptibility to or causes a disease. Gene identification comes next once an SNP relationship has been verified across large populations. The candidate genes with known activities that explain the disease may be found in the human genome sequence close to the SNPs. Genome-wide association studies are also used to explore typical traits like height and body mass index. Twenty lip descriptors, eye and nose positions, and the distance between the nose and top lip were all measured by researchers. The investigation discovered variations of various genes that are already known to be mutant in specific diseases that impair facial feature development. A genome-wide association research can look into the multiple facets of characteristics and diseases. Another GWAS is looking for smoking-related behaviors that could point to a higher chance of genetic susceptibility leading to lung cancer in humans. These potentially instructive behaviors include the age at which smoking started, the length of time a person has smoked, the length of time since they stopped, and the number of cigarettes smoked each day. As the activities of genes are found, genome-wide association studies are being superseded by direct examination of the human genome sequence due to their shortcomings. Genome-wide association studies have the conceptual drawback of revealing relationships between sets of data rather than underlying causes. An association simply denotes the occurrence of one event or trait in conjunction with another. When two measurements increase together, such as when stress and blood pressure do, there is a link. A detailed mechanism explaining how one event causes another is necessary in order to prove a reason, for example, how does stress increase blood pressure? An association study is more of a tool for discovery than it is a source of knowledge about a gene's function.

The fact that GWAS studies frequently find regions of the genome that just marginally increase the probability of contracting a disease is a practical drawback of these investigations. To a smoker whose environmental risk is significantly higher, a genetic test that shows a 1% increase in the risk of developing cancer, for example, would not matter much. Even if a gene, for instance, contributes so little to the risk of cancer that it is missed by a

GWAS, it may still be the cause of the disease in a certain family. A GWAS research may contain bias due to how a patient population was chosen. For instance, persons who died or are only very minimally afflicted would not be included in samples taken from clinics. The fact that the control population may not truly be made up of healthy people is another potential cause of inaccuracy. Other than the issue under investigation, they can have other issues. The accuracy of genome-wide association studies is further impacted by the complicated issues. Remember that a phenocopy is an ailment or trait that resembles an inherited one but is caused by the environment. It would be deceptive to include persons who have inherited anemia with those who have anemia brought on by a drug reaction. Another potential source of inaccuracy is genetic heterogeneity, in which several genes can result in the same trait or disease. These studies are also complicated by epistasis, which occurs when one gene obscures the impact of another; nevertheless, as we understand more, such interactions are being taken into consideration. Another drawback of genome-wide association studies is the possibility that individuals who have symptoms and an SNP pattern may also have additional factors in common that also contribute to the link, such as a history of environmental exposure, which could lead to a false-positive result. For instance, mutations raise the risk of atherosclerosis, but so do infections, smoking, inactivity, and a diet high in fat. These environmental influences are so widespread that a GWAS may fail to appropriately identify a genetic influence if it is not large enough.

The caliber of the research question asked may determine whether a GWAS is successful. For instance, the method proved particularly helpful in explaining why some Solomon Islands residents have blond hair. The majority of the inhabitants of these equatorial islands have dark complexion and hair, just like those who reside in equatorial Africa. The blonds were significantly more likely to have a specific SNP on chromosome 9 than the dark haired islanders, according to a case-control genome-wide association research on 43 blond and 42 dark haired Solomon Islanders. All vertebrate animals' melanin pigmentation is regulated by its protein product, and the defective gene it produces results in a form of albinism in humans. Some Solomon Islanders' distinctive blond hair is caused by a single DNA nucleotide mutation. The old and modern methods for analyzing multifactorial features frequently complement one another. The same is true of stuttering. For MZ twins, concordance ranges from 20 to 83 percent, whereas for DZ twins, concordance ranges from 4 to 9 percent, indicating a significant genetic component. An analysis of 100 families with at least two stutterers revealed candidate genes on three chromosomes that may be involved in the characteristic.

5.9 UNDERSTANDING THE APPETITE

Eating normally causes fat cells (adipocytes) to secrete leptin, which circulates via the blood to the hypothalamus of the brain and binds to receptors on nerve cells (neurons). In response to leptin binding, neurons produce a different hormone that attaches to yet more receptor types and, in turn, acts as an appetite "brake" while accelerating the digestion of previously consumed food. Leptin levels drop after many hours without food, which causes the release of an appetite "accelerator." There is a lot of interest in using medications to target genes and proteins that regulate appetite in order to gain weight or lose it. Friedman observed that mice who received more leptin ate less and lost weight.

A peptide hormone called ghrelin, which is released in the stomach in response to hunger, tells the hypothalamus to make more of the appetite stimulant (Yang, Brown, Liang, Grishin, & Goldstein, 2008)appetite-stimulating peptide hormone secreted by the food-deprived stomach. Serine-3 of ghrelin is acylated with an eight-carbon fatty acid, octanoate, which is required for its endocrine actions. Here, we identify GOAT (Ghrelin O-Acyltransferase. Making the stomach smaller is one method that weight loss surgery might reduce ghrelin release. The hormones in the stomach that regulate appetite work in the short term, whereas leptin functions long-term to maintain weight. The combination of all these hormonal signals allows for precise appetite regulation that keeps weight stable. The discovery of specific genes that affect weight opened the door to the multifactorial theory of the attribute. Researchers are looking into how different gene combinations affect weight. One study examined 21 genes whose mutations result in disorders that include obesity as well as 37 genes whose products are involved in weight-related biochemical processes. This method discovered many unusual gene variants that, when combined, may explain why many people tend to put on weight. Understanding of body weight has been improved as a result of genome-wide association studies that contrast gene expression patterns. In one study, the gene expression patterns in adipose (fat) tissue and other tissues were compared. More than 1,600 Icelanders' blood samples were used to identify a group of genes whose byproducts contribute to immunological function, inflammation, and obesity-related features.

The results of many studies on twins and adopted people point to a heritability of 75% for obesity. The gap shows that genes play a larger influence in those who tend to acquire weight easily because the heritability for BMI is lower than this. When populations with a genetic propensity to readily gain weight face a significant and abrupt decline in the quality of the diet, the genetic component of obesity can be detected. When the people of the little island of Naura in Western Samoa discovered a market for the massive amounts of bird waste on their island as commercial fertilizer, their way of life drastically changed. The money replaced an agrarian lifestyle and diet of fish and vegetables with inactivity and a high-calorie, high-fat diet. Two-thirds of the population had gained weight within a generation, and one-third developed type 2 diabetes.

Another example of how the environment can affect body weight comes from the Pima Indians. This population split into two populations, with one part settling in Mexico's Sierra Madre mountains and the other in southern Arizona. The Arizona Indians stopped farming and eating a low-calorie, low-fat diet by the 1970s and started getting 40% of their calories from fat. They saw the greatest prevalence of obesity of any population on earth as a result of this radical change in lifestyle. By the time they turned 35, half of the Arizona group had developed diabetes and were, on average, 57 pounds heavier than their low-fat-eating, highly active southern counterparts. The Pima Indians show that future obesity is not predetermined by genes at conception but is far more likely to happen if the environment contains an excessive amount of calories and fat. Geneticist James Neel in 1962 proposed that the hunter-gatherers who survived the famine in the distant past had genes that allowed them to store fat (Fee, 2006). Given the abundance of food available now, the genetic propensity to store fat is detrimental rather than beneficial.

The hypothesized gene also holds true for newborns who were very underweight despite being born after a full-term pregnancy. Before birth, the metabolism changes in a way that conserves calories to counteract starving conditions in the uterus; as a result, the person later confronts an increased risk of heart disease, stroke, obesity, osteoporosis, and type 2 diabetes. These complex situations are the result of epigenetic changes in gene expression rather than mutations. The gut microbiome, or the sorts of bacteria that ordinarily reside in human digestive tracts, is another environmental factor that affects weight. The amount of calories that we derive from particular diets is influenced by the actions of specific types of bacteria. A person who struggles to maintain weight loss has a different gut microbiome than someone who does. Finding a means to reproduce this altered microbiome could one day offer a surgical alternative to the drastic changes in the gut flora that occur after weight-loss surgery. The diseases that are a result of gene-environment interactions may be the areas where these complexity and difficulties are most significant.

5.10 PTSD

An adult lady who was physically and psychologically tortured by her uncle as a child grows up to be a loving mother and achieves success in her work. Another woman who had a comparable background developed into a fearful adult with posttraumatic stress disorder (PTSD). Traumatic physical harm or the perceived fear of harm can cause PTSD, which lasts for a long time after the first triggering event. Because the stress hormones that flooded the bloodstream during the initial trauma were improperly activated, the person experiences anxiety, flashbacks, and avoidance of recollections of the event. PTSD is frequently linked to military service, although it can also occur after medical emergencies, catastrophes, criminal activity, and other upsetting occurrences.

According to a study on PTSD (Figure 5.5) brought on by childhood abuse, a mutation of one specific gene may have a substantial impact on why some victims of childhood abuse go on to acquire PTSD as adults and others do not (Heide & Solomon, 2006).

Researchers looked at variations of the *FKBP5* gene in 2,000 people who had experienced severe abuse as either children or adults. A specific version of the gene was substantially more prevalent in those with PTSD who had experienced maltreatment as children than in the general population. The outcome has genetic and epigenetic components. Methyl groups on the gene are deleted in individuals with the gene variant who experienced maltreatment as children and went on to develop PTSD; this impact is not present in individuals who did not. The body is unable to respond to stress and adapt as a result of the decreased methylation, which causes stress hormone (glucocorticoids) receptors to fail. The epigenetic modification was discovered in white blood cells, but subsequent laboratory cell growth studies revealed that it also happens and remains in progenitor cells that differentiate into lifelong brain nerve cells. The epigenetic shift occurs in a person at a crucial point of development, preparing a child for a later inability to adequately utilize stress hormones. PTSD might be the outcome.

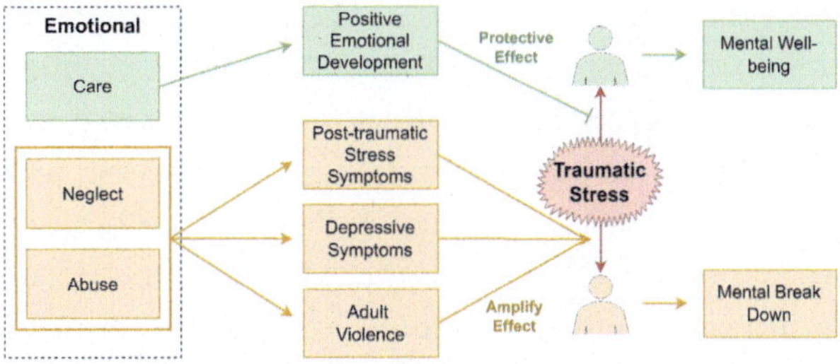

Figure 5.5. PTSD. (Source: Al Jowf GI, Ahmed ZT, An N, Reijnders RA, Ambrosino E, Rutten BPF, de Nijs L, Eijssen LMT. A Public Health Perspective of Post-Traumatic Stress Disorder. *International Journal of Environmental Research and Public Health*. 2022; 19(11):6474. https://doi.org/10.3390/ijerph19116474, Creative Commons License).

How we behave on a daily basis is governed by a complicated continuum of emotions, moods, intelligence, and personality. To some extent, our many behaviors define and assess us. They determine how we interact, deal with unpleasant emotions, and handle stress. The symptoms of behavioral disorders can be varied and occasionally overlap. Developing treatments for behavior-based diseases can be aided by understanding the biology that underlies behavior. The study of behavioral genetics takes into account the function and diversity of the neurological system, as well as the elusive traits of mood and thought. The human brain weighs 3 pounds and looks like a large gray walnut. It is made up of 100 billion neurons, the building blocks of the nervous system, and at least a trillion neuroglia, or supporting cells, which help the neurons grow and function. Complex networks of interconnected brain neurons govern the entire body. Each of the 100 billion neurons has branches that create intimate connections with 1,000 to 10,000 other neurons. These connections are called synapses.

Using chemical signals known as neurotransmitters, neurons communicate with one another across these tiny spaces. Networked neurons control a wide range of processes, including perception and sensation, memory, thought, and muscle action. By regulating the synthesis and distribution of neurotransmitters, genes have an impact on behavior. Neurotransmitters are synthesized by enzymes and transported from the transmitting (presynaptic) neuron to receptors on the receiving (postsynaptic) neuron across the synapse. Neurotransmitters are transported by proteins known as transporters, and proteins also make up the receptor subunits. Myelin, a coating made of proteins and lipids that covers axons of neurons, is produced under the direction of genes. The neuron is protected by myelin, which quickens neurotransmission. Like other qualities, many behavioral traits and diseases are likely the result of a single gene, maybe one whose protein product directly participates in neurotransmission, as well as minor contributions from frequent gene variants. Researchers predict that between 100 and 300 genes are active. An individual is more likely to experience the onset of a specific condition if they inherit a specific group of variants of these genes in the presence of a specific environmental trigger. More than one behavioral problem has been linked to some genes. Traditional empiric risk, adoption, and twin research data as well as, increasingly, exome and genome sequencing are used to find gene variants related to behavioral traits and disorders in order to decipher the genetic components of behavioral traits and disorders. This method might make it possible to physiologically meaningfully subtype behavioral problems, allowing for more individualized diagnoses and quicker implementation of effective therapies.

5.11 SLEEP AND GENES

Sleep has been said to be a critical activity of unknown function, yet animals do actually perish without it. We reside in this enigmatic state for one-third of our lives (Trouche, Pompili, & Girardeau, 2020). Sleep characteristics are influenced by genes. MZ twins report having substantially more in common than DZ twins when it comes to sleep duration, scheduling, quality, napping habits, and whether they are night owls or morning persons. Even MZ twins who were split up at birth can attest to this. A genetic factor is confirmed by twin investigations of brain wave patterns throughout four of the five stages of sleep. The fifth stage of sleep, known as REM sleep, is linked to dreaming and may thus be more influenced by experience than by heredity. Narcolepsy is a genetic disorder that affects people. Narcolepsy causes a human to wake up unexpectedly throughout the day. Extreme daytime sleepiness seriously interferes with daily tasks. The frequency of automobile accidents is five times higher in narcoleptics. The inability to move for a short while after awakening is known as sleep paralysis. Cataplexy is the most severe narcolepsy symptom. The jaw sinks, the head descends, the knees buckle, and the person collapses to the ground during these brief and abrupt episodes of muscle weakness. This frequently happens when someone is laughing or feeling excited. Only 0.02 to 0.06 percent of people in North America and Europe suffer from narcolepsy with cataplexy, but the fact that it is substantially more prevalent in some families raises the possibility of a genetic cause. One human narcolepsy gene was discovered as a result of research on dogs. In 1999, scientists identified changes in a gene that creates a receptor for the neuropeptide hypocretin. In Doberman and Labrador retrievers, the receptor does not reach the cell surfaces of certain brain cells, preventing the cells from receiving signals to promote a state of awareness.

The suprachiasmatic nuclei are two clusters of neurons in the brain that function as a circadian pacemaker, regulating daily rhythms like the sleep-wake cycle. Certain clock genes in these cells are activated in response to environmental light or darkness. In families where there is a mutation, the function of clock genes is astonishing. For instance, a family from Utah with familial advanced sleep phase syndrome spans five generations. Affected individuals go to bed at 7:30 every night and abruptly awaken at 4:30 in the morning. They have a single DNA base mutation in the gene, which is on chromosome 2, which prevents them from synchronizing their sleep-wake cycle with the daily sunrise and sunset, an example of a circadian rhythm. How soon we typically awaken depends on a mutation in a different location of the gene. Genotype AA individuals who have two adenines at this location awaken an hour earlier than individuals who have two guanines. When the two homozygotes are sleeping, heterozygotes of genotype AG awaken.

A complicated and subjective attribute, intelligence is influenced significantly by both inherited and environmental factors. Many definitions exist for intelligence. The ability to reason, learn, recall, link concepts, deduce, and create are all examples of intelligence. In the late nineteenth century, the first intelligence tests were created to measure a person's sensory perception and response times to diverse stimuli. In order to predict the academic achievement of children with developmental disabilities, Alfred Binet in France created a test in 1904 that included questions based on language, mathematics, and drawings. At Stanford University, the test was altered to evaluate white, middle-class Americans.

5.12 BRAIN CUES AND ADDICTION

Tolerance and reliance are the two distinguishing traits. As time goes on, tolerance causes the need to take more of the drug to have the same results. When a drug user stops using it, they develop withdrawal symptoms, which is known as dependence. The biological/psychological aspects of drug addiction are influenced by both tolerance and dependence (Buccelli et al., 2016)use patterns, acceleration of disease course, and help-seeking patterns are known to be influenced by gender differences with regard to biological, psychological, cultural and socioeconomic factors. This paper presents a systematic review of published data on gender differences in the use/abuse of psychoactive and psychotic drugs, focusing on the importance of a multidisciplinary approach. The basis for this paper was obtained by Medline searches using the search terms "human" and "gender", combined with individual drug names or "drugs of abuse". The reference lists of these papers were further checked for other relevant studies. The gender difference in drug abuse is more evident in adults than in adolescents (13–19 years. It can be quite challenging to stop drug addiction-related behavior.

Brain alterations brought on by drug addiction are permanent. Even after years of abstinence, cravings and a significant risk of relapse persist. Heritability ranges from 0.4 to 0.6, with an increased risk of two to three times for adopted people with one affected biological parent. Twin studies suggest that drug addiction has an inherited component. By highlighting the cell surface receptors that bind neurotransmitters when a person craves the drug, brain imaging techniques can pinpoint where the site of drug addiction is located in the brain. The limbic system, which includes the nucleus accumbens, prefrontal cortex, and ventral tegmental region, is a bigger group of brain structures where the changes in the brain that cause addiction take place. Alcohol impacts the prefrontal cortex, but cocaine appears to mostly have an effect on the nucleus accumbens. The ability to feel good after consuming certain natural chemicals, especially those derived from plants, must have been known to our distant ancestors. This inclination still exists now. Chemicals originating from plants are frequently used as drugs; examples include cocaine, opium, and tetrahydrocannabinol (THC), the primary psychoactive component of marijuana. These compounds bind to receptors on human neurons, suggesting that our bodies naturally produce these compounds. Endorphins, enkephalins, and anandamide are the human analogs of opiates and THC, respectively. Enkephalins and endorphins reduce pain. Anandamide binds to neurotransmitter receptors on presynaptic (sending) neurons, altering how brain cells react to stimuli. Neurotransmitters, on the other hand, attach to receptors on postsynaptic neurons. Both amphetamines and LSD work by attaching to neuronal receptors that typically bind trace amine neurotransmitters, which are present in small amounts throughout the brain. LSD has effects that are comparable to schizophrenic symptoms, which raises the possibility that the trace amine receptors, which are proteins, are connected to the disease. There are specific gene variations that are common among drug addicts that need to be combined with environmental cues for addiction to take place. For instance, individuals with alcoholism and other addictions are overrepresented among those who are homozygous for the dopamine D(2) receptor gene variant's A1 allele. More than 50 chromosomal regions have been identified by genome-wide association analyses as potentially harboring genes involved with desire. Gene expression research adds further detail to this picture by allowing researchers to see in real-time the biochemical changes that take place when a person seeks a drug and then consumes it. Finding the genetic roots of nicotine addiction advances our understanding of addiction in general and could have useful applications (Merikangas & McClair, 2012). Genetics may help to explain why nicotine is so extremely addictive and how nicotine causes lung cancer.

5.13 DEPRESSION AND GENETICS

Clinical depression is characterized by a range of incapacitating symptoms. A depressed person no longer takes pleasure. He or she is worn out and struggles to focus, make judgments, and remember specifics. Anxiety and impatience increase, motivation suffers, and crying may start suddenly, frequently multiple times per day. Depression makes it harder for people to fall asleep and stay asleep, which makes them feel more tired during the day. Simple, everyday chores could become overwhelming, making it very challenging to work or study. Weight loss may occur as interest in food, like most other things, decreases. Depression has almost equal contributions from the environment and genes. The risk for the general population is 10%, but it is 20% to 30% for those who have a first-degree relative who is affected. Although many of the genes associated with depression are likely still to be discovered, by concentrating on sleep disruption, researchers were able to identify some. Eight of the genes they searched for that are known to regulate circadian rhythms, sleep, hormone levels, and body temperature showed altered expression in depressive individuals' brains when compared to those of non-depressed individuals. The neurotransmitter serotonin deficit in synapses is probably a major contributor to depression. It has an impact on feelings, appetite, sleep, and mood. SSRIs, which are antidepressant medications, stop presynaptic neurons from absorbing serotonin from the synapse, so increasing the amount of it accessible to excite the postsynaptic cell.

This behavior seems to make up for the neurotransmitter deficit. The serotonin transporter protein is connected to depression genetically because it may operate too well in depressed people and remove excessive amounts of serotonin from the synapses. A person with moderate or severe depression may be able to resume some activities after one week because of the effects of SSRIs, but a full recovery may take up to eight weeks. In the past, doctors would test one antidepressant medication after another to treat patients who had depression based on their prior patient experiences. This method of trial and error would frequently take months. Researchers are discovering profiles of gene variations or gene expression that are connected to a particular drug's reaction as the genes in the human genome are identified. In the future, doctors will be able to choose the medication that will help a patient beat depression the fastest and with the fewest side effects.

Depression is significantly more common than bipolar disease. A lifetime risk of 0.5 to 1.0 percent for the general population applies, and it affects 1% of the population. Hypomanic bipolar disease can mimic depressive symptoms. Due to the fact that different medications are used to treat depression and bipolar disorder, this is a crucial clinical distinction. Bipolar disorder is influenced by several gene variations. Large Amish families were the focus of early genetic investigations because their austere way of living made the overexcited phase of the condition highly visible in them. However, research in other families linked several genes. There are various gene variant combinations that either cause or contribute to the phenotype of bipolar disorder, but only a handful of these variants are present in any one family, which causes studies to differ (Hennah et al., 2009)schizoaffective disorder, bipolar disorder, major depression, autism and Asperger syndrome, but apart from in the original translocation family, true causal variants have yet to be confirmed.

Here we report a harmonized association study for DISC1 in European cohorts of schizophrenia and bipolar disorder. We identify regions of significant association, demonstrate allele frequency heterogeneity and provide preliminary evidence for modifying interplay between variants. Whereas no associations survived permutation analysis in the combined data set, significant corrected associations were observed for bipolar disorder at rs1538979 in the Finnish cohorts (uncorrected P=0.00020; corrected P=0.016; odds ratio=2.73±95% confidence interval (CI.

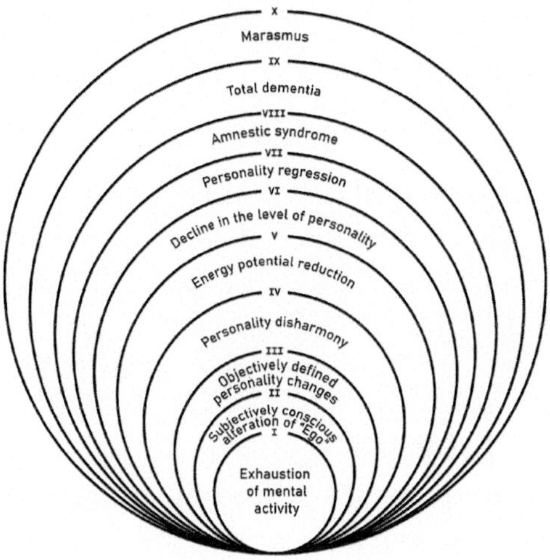

Figure 5.6. Schizophrenia and its negative symptoms. (Source: Mosolov SN and Yaltonskaya PA (2022) Primary and Secondary Negative Symptoms in Schizophrenia. Front. Psychiatry 12:766692. doi: 10.3389/fpsyt.2021.766692, Creative Commons License).

5.14 SCHIZOPHRENIA

Schizophrenia (Figure 5.6) is characterized by a loss of mental flexibility that causes disengagement from reality. It is a type of psychosis, which is a mental condition and loss of identity. The dementias are cognitive, while the mood disorders are affective. A total of 1% of people worldwide are affected by schizophrenia in some way. Ten percent of those impacted commit suicide. The challenges in studying a behavioral disease are highlighted by the identification of genetic components to schizophrenia. Thinking is frequently impacted by schizophrenia's initial symptoms. Learning may become challenging as a person's memory deteriorates and their capacity to retain knowledge decreases in late childhood or early adolescence. Between the ages of 20 and 37 for women and 17 to 27 for men, psychotic symptoms usually appear. Delusions and hallucinations come under this category. Schizophrenia patients may detect a voice offering directions. Schizophrenia patients have extremely real uncertainties that most people might consider unreasonable, such as being pursued by monsters. Speech mimics the jumbled cognitive process, jumping from subject to subject with no discernible rationale. Inappropriate reactions could include laughing at depressing news. A person with schizophrenia may express their mental disintegration through their artwork. Despite its name, schizophrenia does not lead to numerous or divided personalities. The progression of schizophrenia often reaches a peak or becomes episodic in these patients. Unlike dementia, there is no continual decline with this condition.

Living with and copying those who have schizophrenia might cause some symptoms, such as disorganized thinking. There is also a strong environmental component to the disease. Schizophrenia may be brought on by the interaction of dozens of genes with an environmental trigger or triggers. A minimum of twenty-four sites where affected siblings share alleles are found in families with schizophrenia, which is far more frequently than the 50% that Mendel's first law predicts. These areas might include genes that support the growth of schizophrenia. Rare DNA duplications or deletions that develop in individuals with schizophrenia are particularly likely to do so because they are acquired rather than inherited.

The chance of having schizophrenia may be increased by a number of environmental factors. These include difficulties during childbirth, low oxygen levels in the fetus, herpesvirus infection at birth, and maternal starvation or traumatic brain injury. A particularly well-studied environmental factor that increases the incidence of schizophrenia is infection during pregnancy (A. S. Brown, 2012). The immune system of infected pregnant women floods the embryonal brain with cytokines, which cause inflammation and may gradually affect brain development. Since many viruses cannot pass through the placenta and reach the fetus, it is most likely the mother's immune response that raises the risk. Observations that many persons with schizophrenia were born in the winter, particularly during flu pandemic years, led to the hypothesis that maternal infection may play a role in the development of schizophrenia. This observation is supported by research. There is a link between having a kid who develops schizophrenia and having high levels of the cytokine interleukin-8, according to blood samples taken from 12,000 pregnant women in the 1950s and 1960s.

After birth, environmental factors may increase a person's risk of developing schizophrenia. Four of these risk factors were found in a study of 7,500 persons with schizophrenia conducted in Europe: living in a city, migrating, regularly using marijuana, and being a victim of abuse, neglect, or bullying as a child. Starvation is another environmental factor that can significantly raise the risk of schizophrenia. One class of schizophrenia-causing mutant genes offers a possible explanation for how the disease might develop. These genes produce methyltransferases, which are enzymes that transfer methyl groups to protect specific genes at particular periods, suppressing the expression of those genes. The fact that the DNA sequences are unaltered makes this control of gene expression epigenetic. Researchers evaluated the methyltransferase activity in neurons from the prefrontal cortex region of the brains of people who had died at various ages, from embryos to 80-year-olds, from non-psychiatric causes. More genes are silenced as we become older. Addiction is influenced by the prefrontal cortex, which controls executive function, or the capacity to organize one's thoughts, learn new things, recall the past, and make connections between it and the present. Methyltransferases in the prefrontal cortex excessively silence genes in schizophrenia. The signs of abnormal thinking then take years to occur. Some bipolar disease sufferers also have overactive methyltransferase.

5.15 AUTISM

Even though it is only now being realized, signs of autism disorder that affect socializing and communication abilities start to show before the age of three. As they get older, about 25% of the affected youngsters develop seizures. Even though intellectual handicap affects 70% of those with autism, some may be quite clever. An infant with severe autism may lose language terms and stop initiating discussions. He or she quits playing with other kids and instead prefers to be by themselves. Some youngsters find comfort in repeating actions, rocking back and forth, or grasping a prized object. An individual with autism can connect with the outside world with the use of special education programs that follow a strict schedule. Autism may have socialization consequences but not linguistic problems in its sufferers (Ballaban-Gil, Rapin, Tuchman, & Shinnar, 1996). This condition was formerly referred to as Asperger syndrome.

Various genes, in combination with environmental factors, all have a role in autism to varying degrees. Prenatal rubella (German measles) exposure and valproate use are examples of environmental triggers. Research suggests that folic acid supplementation late in pregnancy may raise the risk of autism by altering the epigenome. Different genetic subtypes may be taken into account in order to compare individuals with the same underlying issue in research, which may help to better uncover environmental risk factors. About 90% of autism cases are heritable. Although there is a 15% chance that siblings of affected children would also have the disorder, compared to a less than 1% chance in the general population, in certain MZ (identical) twin pairs, one twin has autism and the other does not. More than 30 genes have been shown to be causal or susceptible. A syndrome that includes single-gene diseases like Rett syndrome, chromosomal disorders like fragile X syndrome, and Down syndrome affects roughly 15% of individuals with autism. Testing for these diseases and a test called chromosomal microarray analysis, which finds copy number variants, are all part of the diagnosis process for autism. The advancement of genetic technology has coincided with our understanding of the genetics of autism. Researchers are now able to pinpoint many more of the genes responsible for autism because of whole-genome and exome sequencing initiatives. Within the enormous datasets that these technologies provide, experiments concentrate on candidate genes.

Chapter

6 MEDICAL GENETICS

CONTENTS

6.1 INTRODUCTION

Any use of genetics in medicine is referred to as medical genetics (Kearney, Thorland, Brown, Quintero-Rivera, & South, 2011). Thus, it covers research on how diseases are passed down through families, mapping of disease genes to specific chromosomal sites, study of the molecular mechanisms by which genes cause disease, as well as the diagnosis and treatment of genetic disease. Rapid advancements in molecular genetics have made it possible to diagnose thousands of inherited diseases using DNA, and gene therapy, which involves inserting healthy genes into patients to treat genetic disease, is showing promise for some conditions.

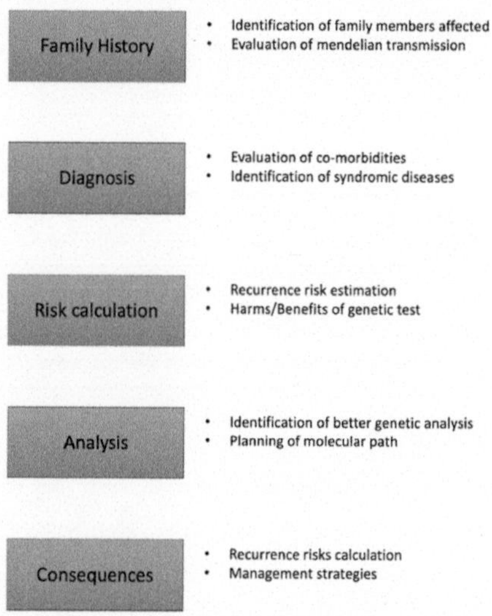

Figure 6.1. Genetic counseling. (Source: Zampatti S, Ragazzo M, Peconi C, Luciano S, Gambardella S, Caputo V, Strafella C, Cascella R, Caltagirone C, Giardina E. Genetic Counselling Improves the Molecular Characterisation of Dementing Disorders. *Journal of Personalized Medicine*. 2021; 11(6):474. https://doi.org/10.3390/jpm11060474, Creative Commons License).

Genetic counseling, (Figure 6.1) which informs patients and their families of risks, prognoses, and therapies, is another aspect of medical genetics. For thousands of years, people have been curious about and interested in the inheritance of physical characteristics. Many genetic occurrences were described by ancient Hebrews and Greeks as well as later medieval scholars who put out ideas to explain them. Many of these theories turned out to be false. The father of genetics, the Austrian monk Gregor Mendel, made substantial contributions to science by carrying out a number of skillfully planned tests on living things (garden peas). He then developed a number of key theories of heredity using the experimental data. In 1865, Mendel published the findings of his research in a publication that was comparatively inconspicuous. One of the ironies of biological science is that for 35 years, little attention was paid to his discoveries, which are today the foundation of genetics. The development of genetics as we know it now was mainly influenced by research done in the 20th century. In 1900, three distinct scientists working in three different countries independently discovered Mendel's concepts. Landsteiner also proposed the ABO blood group this year. The first "inborn error of metabolism" was alkaptonuria, according to Archibald Garrod in 1902 (Bearn, 1996). Johannsen first used the term gene to describe the fundamental component of inheritance in 1909.

There was a lot of experimental and theoretical work done over the course of the following several decades. *Drosophila melanogaster* (fruit flies) and *Neurospora crassa* (bread mold) were two examples of organisms that made good experimental platforms for examining the functions and interactions of genes. H. J. Muller, for instance, used the fruit fly as a model to show the genetic effects of ionizing radiation. During this time, three key scientists, Ronald Fisher, J. B. S. Haldane, and Sewall Wright developed a large portion of the theoretical foundations of population genetics. A number of genetic disorders, including phenylketonuria, sickle cell anemia, Huntington's disease, and cystic fibrosis were studied. Oswald Avery demonstrated in 1944 that deoxyribonucleic acid makes up genes. The description of the physical structure of DNA by James Watson and Francis Crick in 1953 was arguably the most important discovery of the 1950s. Correctly calculating the number of human chromosomes was another key achievement of that decade. It had been believed that each human cell contains 48 chromosomes since the early 1920s. The precise number, 46, was only discovered in 1956. A surge of new discoveries in cytogenetics resulted from the capacity to count and identify chromosomes, including the discovery that an extra copy of chromosome 21 causes Down syndrome in 1959. Since 1960, substantial advancements in technology have occurred at an ever-increasing rate. Many chromosomal sites have been assigned to thousands of genes. The Human Genome Project, an endeavor that started in 1990, produced a full sequence of human DNA in 2003. The massive amount of data being produced by this research and others like it has been deciphered due to significant advancements in computer technology. Geneticists have identified the molecular abnormalities behind thousands of genetic diseases in addition to mapping genes. Our knowledge of how gene abnormalities can cause disease has tremendously benefited from this study, paving the door for new treatments and cures.

6.2 GENETIC DISORDERS

According to estimates, humans have between 20,000 and 25,000 genes. Although many genetic disorders are the consequence of a combination of genetic and nongenetic variables, some genetic disorders, notably the single gene syndromes, are strongly governed by genes. Therefore, one might think of genetic diseases as existing along a continuum, with problems like measles located at one end (highly determined by environment) and disorders like cystic fibrosis and Duchenne muscular dystrophy situated at the other end (strongly determined by genes). Many of the most common conditions, such as birth abnormalities and widespread diseases like diabetes, hypertension, heart disease, and cancer, fall somewhere in the middle of the spectrum. These diseases are the result of genetic and environmental effects to varying degrees. Diseases with mostly nongenetic causes (i.e., those brought on by starvation, unhygienic surroundings, and infections) accounted for the majority of child fatalities less than a century ago. Public health did, however, significantly advance during the 20th century. Genetic diseases account for a sizable portion of childhood fatalities in addition to a sizable portion of admissions to pediatric hospitals. For instance, a survey of pediatric inpatients at Seattle hospitals revealed that 27% of all patients had a genetic disease, while a survey of admissions to a prominent pediatric hospital in Mexico revealed that 37.8% of cases had a disease that was either genetic or partly genetic (Ashbolt, 2004)although poor sanitation and food sources are integral to enteric pathogen exposure. Gastrointestinal disease outcomes are also more severe, due to under-nutrition and lack of intervention strategies in these regions. Poor water quality, sanitation and hygiene account for some 1.7 million deaths a year world-wide (3.1% of all deaths and 3.7% of all DALY's.

Cystic fibrosis is more prevalent in people of European ancestry, but sickle cell disease is more prevalent in those of African ancestry. Some diseases affect elderly people more frequently. For instance, a tiny percentage (5–10%) of diseases including colon cancer, breast cancer, and Alzheimer's are brought on by dominant genes, yet these conditions typically do not show symptoms until later in life. Estimates of these genetic diseases' prevalence in elderly people would be higher. Prevalence estimates may differ due to variations in diagnostic procedures. It is noteworthy that a recognized genetic condition will be identified in at least 3 percent to 7 percent of the population at some point, keeping these sources of variation in mind. Cellular defects are present in all genetic disorders. Because of this, understanding genetic disease requires a basic understanding of cell biology.

Both the replication of genetic information and the translation of genes into proteins are subject to error. Single-gene diseases are frequently caused by such errors. Moreover, mistakes that can take place during cell division can result in chromosome-wide diseases. Scientists first hypothesized that the crucial mechanisms of heredity are located in the nucleus of the cell in the 19th century as a result of microscopic examinations of cells. They discovered that non-dividing cell nuclei can be observed to contain chromatin, the component that gives the nucleus its granular appearance. Chromosomes are threadlike structures that may be seen under a microscope just before a cell divides. Early in the 20th century, Mendel's breeding experiments were rediscovered, and it soon became clear that chromosomes carry genes. The fundamental component of inheritance, the gene, is passed from parent to child. Physical characteristics like eye color are passed down via families via gene inheritance. Furthermore, diseases can be passed down through genetic heredity.

6.3 GENES

Genes are made of deoxyribonucleic acid on a physical level. All proteins in the body have DNA as their genetic blueprint. Thus, genes eventually have an impact on every element of body composition and function. According to estimates, humans have between 20,000 and 25,000 genes (DNA sequences that code for proteins or ribonucleic acid [RNA]). These genes frequently contain errors (or mutations), which frequently result in recognized genetic diseases. There are 46 chromosomes altogether in each human somatic cell (i.e., cells other than sperm, egg, or gametes). There are 23 pairs of distinct chromosomes in each somatic cell. Each pair has two members, one descended from the mother and the other from the individual's father. The sex chromosomes are located on one of the chromosome pairs. The Y chromosome, which is inherited from the father, and the X chromosome, which is inherited from the mother, make up the sex chromosomes in males. In typical females, there are two X chromosomes, one from the father. Autosomes make up the remaining 22 pairs of chromosomes. Somatic cells are diploid because they contain two copies of each chromosome. The haploid number of chromosomes in human gametes is 23. In contrast to meiosis, which produces haploid chromosomes, mitosis allows somatic cells to preserve their diploid number of chromosomes over multiple generations. Cytosine and thymine are two bases that are classified as pyrimidines, which are single carbon-nitrogen rings. The other two bases, adenine and guanine, are purines, which are double carbon-nitrogen rings. The initials C, T, A, and G are frequently used to symbolize the four bases.

A contribution made by Watson and Crick in the middle of the 20th century was to show how these three elements are physically put together to produce DNA. They put forth the famous double helix concept, in which chemical linkages serve as the rungs of a twisted ladder that represents DNA. Strong phosphodiester bonds hold the two sides of the ladder together, which are made up of sugar and phosphate components. The nitrogenous bases emerge at regular intervals from each side of the ladder. A rather weak hydrogen bond connects the base projecting from one side to the base projecting from the other side. The ladder's rungs are thus formed by the pairs of nitrogenous bases. Identical copies of the DNA must be created and incorporated into the new cells as cells divide to create copies of themselves. If DNA is to act as the primary genetic building block, this is crucial. When the weak hydrogen connections between bases break, a single DNA strand with unpaired bases is created.

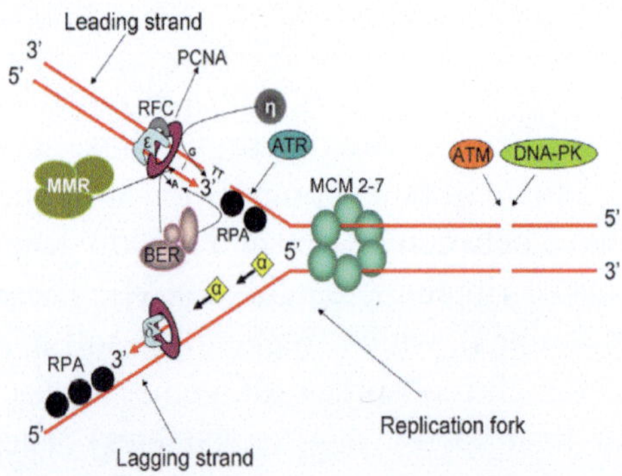

Figure 6.2. DNA replication. (Source: Zannis-Hadjopoulos, M., & Rampakakis, E. (2011). Synergy Between DNA Replication and Repair Mechanisms. InTech. doi: 10.5772/24164, Creative Commons License).

This is when DNA replication starts. Accurate replication depends on the complementary base pairing process, which consistently pairs adenine with thymine and guanine with cytosine. According to the complementary base pairing principle, a free nucleotide will only be drawn to an unpaired base if it has the right complementary base. A section of a single strand, for instance, with the base sequence ATTGCT will bind to a string of free nucleotides with the base sequence TAACGA. The complementary strand is said to be constructed using the single strand as a foundation. A new double-stranded molecule that is identical to the previous one is created after replication is finished.

6.4 REPLICATION

DNA replication (Figure 6.2) involves a number of distinct enzymes. The double helix is unwound by one enzyme, and the strands are held apart by another. Adding free nucleotides to the 3′ end of the new strand is what DNA polymerase, yet another enzyme, does as it move down the single DNA strand. Replication always moves from the 5′ to the 3′ ends because nucleotides can only be added to the 3′ end of the strand. The 5′ direction is referred to as upstream and the 3′ direction is referred to as downstream when discussing the orientation of sequences along a gene. DNA polymerase not only adds new nucleotides but also participates in a proofreading process, ensuring that freshly added nucleotides are in fact complementary to the template base. If not, the nucleotide is removed and is then substituted with the appropriate complementary nucleotide base. DNA replication is significantly more accurate as a result of this mechanism. A mutation happens when a DNA replication fault cannot be successfully corrected. Human DNA replication occurs at a relatively slow rate of 40 to 50 nucleotides per second. Bacteria may move 500–1000 nucleotides per second, which is a far higher rate. Replication would be an incredibly time-consuming operation if it moved linearly from one end of the chromosome to the other, given that some human chromosomes can contain up to 250 million nucleotides. Replication instead starts at a variety of locations along the chromosome known as replication origins. Replication bubbles are the separations of the DNA strands that arise. The replication process might go more swiftly when it takes place concurrently at many locations along the chromosome.

Protein synthesis starts in the cytoplasm of the cell, whereas DNA synthesis and replication occur in the nucleus (Tolonen et al., 2001)a member of the poxvirus family, is unique among most DNA viruses in that its replication occurs in the cytoplasm of the infected host cell. Although this viral process is known to occur in distinct cytoplasmic sites, little is known about its organization and in particular its relation with cellular membranes. The present study shows by electron microscopy (EM. To control the composition of proteins, the genetic information from the DNA must first be delivered to the cytoplasm. There are two steps involved in this: transcription and translation. The DNA code is, in essence, translated into messenger RNA, which exits the nucleus and is then translated into proteins. Ribonucleic acid (RNA), a kind of nucleic acid that is chemically identical to DNA, mediates both transcription and translation. RNA is made up of sugars, phosphate groups, and nitrogenous bases, just like DNA. It differs from DNA in that one of the four bases is uracil rather than thymine, and the sugar is ribose rather than deoxyribose. Due to its structural similarity to thymine, uracil can combine with adenine. Another distinction between RNA and DNA is that, as opposed to DNA, which often has two strands, RNA typically only has one. The process of creating an RNA sequence from a DNA template is called transcription. Messenger RNA is the type of RNA created by the transcription process. One of the RNA polymerase enzymes, RNA polymerase II, binds to a promoter site on the DNA to start the transcription of mRNA. A promoter is a nucleotide sequence that comes just before a gene. The exposed unconnected DNA bases are then revealed after the RNA polymerase separates some of the DNA strands from one another. The template for the mRNA nucleotide sequence is one of the two DNA strands. Along the DNA template strand, RNA polymerase proceeds from 3′ to 5′, building the complementary mRNA strand from 5′ to 3′. The sense strand of DNA, which does not function as a template, and the mRNA have identical nucleotide sequences due to complementary base pairing, with the exception of the replacement of uracil for thymine.

The inclusion of a chemically altered guanine nucleotide caps the developing RNA molecule 5′ ends shortly after RNA synthesis starts. This 5′ cap appears to aid in preventing RNA degradation during synthesis and later aids in identifying the start site for mRNA translation into protein. Until a sequence of bases known as a termination sequence is reached, transcription continues. The 3′ end of the RNA molecule receives a series of 100–200 adenine bases. The poly-A tail structure is thought to play a role in stabilizing the mRNA molecule to prevent degradation when it enters the cytoplasm. The poly-A tail-attached mRNA bases are eventually destroyed, but RNA polymerase often continues to transcribe DNA for several thousand extra bases. Finally, a single transcribed mRNA strand is left when the DNA and RNA polymerase separate from the RNA strand. Some human genes have many promoters that are positioned in various regions of the gene, such as the gene that can cause Duchenne muscular dystrophy (Sussman,

2002). As a result, the gene transcription can begin at several locations, producing somewhat different proteins. As a result, the same gene sequence can be used to code for distinct proteins in other tissues, such as brain tissue and muscular tissue. In some cells of the body, some genes are expressed. These utility genes produce substances needed for cellular metabolism. However, the majority of genes are exclusively expressed at specific times and in certain tissues. As a result, only a tiny portion of genes are actively transcribed in the majority of cells. This specificity explains why, despite nearly all cells having the exact same DNA sequence, there are many diverse cell types producing various protein products. For instance, the low-density lipoprotein receptor genes are transcribed in liver cells and the globin genes are transcribed in the progenitors of red blood cells.

6.5 TRANSCRIPTION

Transcription is facilitated by a wide variety of proteins. Some of these are referred to as general transcription factors since they are necessary for the transcription of all genes. Others, known as specific transcription factors, play more specialized roles and exclusively activate specific genes at certain developmental stages. The previously mentioned RNA polymerase II is a crucial component of transcription. Despite the fact that this enzyme is crucial for starting transcription by attaching to the promoter region, it is unable to identify the promoter region on its own. Furthermore, it is unable to generate substantial amounts of mRNA on its own. A combination of about 50 distinct proteins must interact for transcription to be effective. General transcription factors fall under this category. These bind to RNA polymerase and certain DNA sequences in the promoter region respectively. The general transcription factors enable successful transcription by enabling RNA polymerase to attach to the promoter region. Interaction with enhancer sequences, which may be thousands of bases upstream or downstream of the gene, can significantly improve the transcriptional activity of specific genes. Enhancers don't work with genes in a direct way. Instead, they are held together by a group of distinct transcription factors known as activators. Enhancers promote the transcription of specific genes at certain times through interactions with activators, co-activators, the transcription complex, and finally the gene itself. While other DNA sequences, known as silencers, assist in repressing the transcription of genes through a similar series of interactions, enhancers aid to raise the transcriptional activity of genes.

The expression of important genes can be improperly controlled by mutations in enhancer, silencer, or promoter sequences, as well as in genes that code for transcription factors, which can result in genetic disease. The intricacy and number of transcription factors enable precise control over the regulation of gene expression. DNA-binding motifs, which are specific protein structures that enable the transcription-factor protein to firmly fit into a specific region of the DNA double helix, are responsible for achieving this. Gene activity can also be correlated with patterns of chromatin condensation or coiling (Dahm et al., 1998). Histone acetylation, the attaching of acetyl groups to lysine residues in the histones, is a typical characteristic of decondensed, or open, chromatin areas known as euchromatin. The chromatin is decondensed by acetylation of histones, which makes it more accessible to transcription factors by reducing their DNA-binding affinity. Therefore, transcription is active in euchromatin. Contrarily, heterochromatin is typically more compact, transcriptionally inactive, and less acetylated. MicroRNAs (miRNA), which are short RNA molecules with 17–27 nucleotides but no protein-coding ability, can also affect how genes are expressed. Instead, they can bind to and down-regulate certain mRNAs, lowering their expression levels, because they are complementary to particular mRNA sequences. It has been discovered that microRNA variation plays significant roles in a variety of cancers. The DNA template base sequence and the primary mRNA transcript are identical complementary sequences. Nuclear enzymes cut portions of the RNA, and the remaining pieces are then joined to create the functional mRNA that will travel to the cytoplasm. Introns are the excised sequences, whereas exons are the sequences that remain to code for proteins. The mature transcript does not exit the nucleus and travel into the cytoplasm until gene splicing is complete.

6.6 SPLICING

Some genes include alternative splice sites that enable the primary transcript of the same gene to be spliced in several ways, resulting in the production of various protein products from the same gene. Similar to replication errors, faults in gene splicing are a type of mutation that can result in genetic disease. A single base cannot be specific for a specific amino acid because there are 20 distinct amino acids and only 4 possible RNA bases. By creating synthetic nucleotide sequences and letting them control the synthesis of polypeptides in the lab, definitive confirmation that amino acids are determined by these triplets of bases, or codons, was achieved. Three of the 64 potential codons, or stop codons, indicate the end of a gene. All 61 of the remaining codons refer to an amino acid. The majority of amino acids can be designated by more than one codon. Thus, it is claimed that the genetic code has degenerated. Although more than one codon can specify the same amino acid, each codon can only specify one amino acid. The process known as translation uses mRNA as a template to create polypeptides. However, mRNA is unable to bind to amino acids directly. Instead, it engages with transfer RNA (tRNA) molecules, which are RNA strands with a cloverleaf structure and roughly 80 nucleotides. Each tRNA molecule has a location at the 3′ end where a specific amino acid can be bound by a covalent bond. The anticodon, a group of three nucleotides at the opposite end of the cloverleaf, pairs with a suitable codon in the mRNA by complementary base pairing. The produced polypeptide chain then receives the associated amino acid. In this manner, mRNA functions via tRNA to specify the order of amino acids.

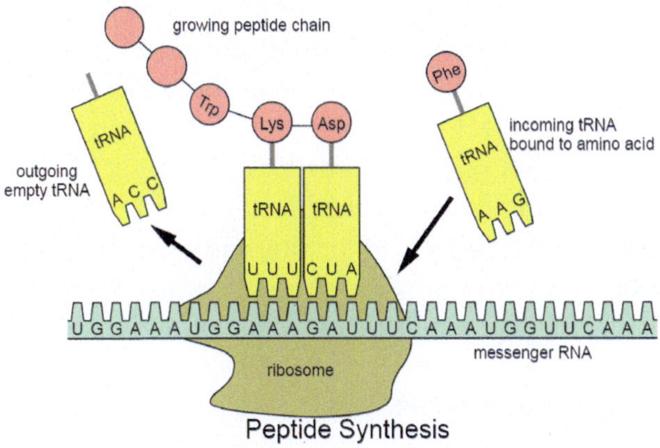

Figure 6.3. Ribosomes play a vital role in translation. (Source: By Boumphrey-fr vector conversion by Glrx - File:Peptide syn.png, CC BY-SA 3.0, https://commons.wikimedia.org/w/index.php?curid=101457889, Creative Commons License).

6.7 RIBOSOME

The ribosome, (Figure 6.3) which is composed of about equal amounts of enzyme proteins and ribosomal RNA, is the cytoplasmic location of protein synthesis (Kudlicki, Coffman, Kramer, & Hardesty, 1997)specifically the 23S rRNA. Refolding and release of enzymatically active rhodanese leaves the ribosomes in an inactive state or conformation for subsequent rounds of refolding. Inactive ribosomes can be activated by elongation factor G (EF-G. rRNA aids in the binding of mRNA and tRNA to the ribosome. The amino acid methionine, which is often eliminated from the polypeptide before polypeptide synthesis is complete, is specified by the codon AUG at this location. Base pairing between the tRNA and mRNA can then take place as a result of the ribosome binding the tRNA to its surface. Codon by codon, the ribosome advances along the mRNA sequence in the 5′ to 3′ direction. The interaction of mRNA and tRNA during the processing of each codon results in the translation of an amino acid. The ribosome acts as a catalyst in this process, supplying an enzyme that causes the covalent peptide bonds to form between the neighboring amino acids, culminating in the growth of a polypeptide. Translation and polypeptide production stop when the ribosome reaches a stop codon in the mRNA sequence. The mRNA strand's 3′ end and 5′ end is represented by the polypeptide's carboxyl (COOH) terminus and amino (NH2) terminus, respectively. The mRNA, ribosome, and polypeptide dissociate from one another after synthesis is finished. After that, the polypeptide is discharged into the cytoplasm. A newly generated polypeptide frequently goes through additional processing, known as post-translational modification, before it can act as a functional protein. In addition to cleaving into smaller polypeptide units, these changes can also combine with other polypeptides to produce larger proteins. The polypeptide could also be altered by having carbohydrate side chains added. These alterations can be required, for instance, to ensure that the mature protein folds correctly or to consolidate its structure. Type I collagen is an illustration of a therapeutically significant protein that experiences significant posttranslational modification.

One feature that sets eukaryotes apart from prokaryotes is the intron-exon structure of genes, which was discovered in 1977. The majority of eukaryotic genes are mostly made up of introns. Enzymes that carry out splicing are directed to the appropriate locations by DNA sequences known as consensus sequences, which are situated adjacent to each exon. Introns are spliced out of the mRNA before it leaves the nucleus, and this splicing must be under control. One interesting theory is that introns, which extend genes, promote gene reorganization during the meiotic exchange of material. Moreover, it has been proposed that introns developed to alter how long DNA replication and transcription take. Only 1% to 2% of the 3 billion nucleotide pairs in the human genome actually function as protein-encoding DNA, despite the fact that this DNA receives the majority of attention in genetics. Most of mRNA functions are unclear, but it has recently been estimated that roughly 75 percent of human DNA is transcribed into mRNA at some point in some cells. Single-copy DNA is the first and most important category of DNA. Single-copy DNA sequences, as the name suggests, are found just once throughout the genome. The genes that code for proteins are found in single-copy DNA, which makes up about half of the genome. The other half of the genome is made up of repetitive DNA, or sequences that occur thousands of times throughout the genome. Satellite DNA and scattered repetitive DNA are the two main categories of repetitive DNA (Warburton & Willard, 1990)the long-range organization of α-satellite DNA from human chromosome 17 was investigated. Three individual chromosomes, representing different α-satellite haplotypes, were segregated into mouse and human somatic cell hybrids and their arrays sized by pulse-field gel electrophoresis. An inventory of the higher-order repeat units found in multiple separate regions of these megabase arrays was obtained using cosmid mapping and two-dimensional gel electrophoresis, a technique that combines the large-scale resolution of pulsed-field gel electrophoresis with the small-scale resolution of conventional gel electrophoresis. These analyses show that α-satellite arrays are characterized by the presence of localized homogeneous domains containing only one distinct type of repeat unit. These domains, which consist of sequence variants and/or higher-order repeat length variants, can be up to at least several hundred thousands of bases in length. Both abundant and rare variant repeat units can be localized in these distinct domains, which may correspond to transition states in the evolution of tandem multicopy DNA families. This description of the organization of large arrays of tandem repeats provides insight into mechanisms involved in their homogenization.”,”container-title”:”Journal of Molecular Biology”,”DOI”:”10.1016/S0022-2836(05. Satellite repeats are grouped together on specific chromosome regions where they are tandem repeats, meaning that the start of one repeat is right next to the end of another. Because they do not occur in tandem, dispersed repeats, as their name suggests, typically exist alone throughout the genome.

6.8 SATELLITE DNA

About 8% to 10% of the genome is made up of satellite DNA, which is further classified into a number of types. Tandem repeats of a 171-bp sequence that can span several million base pairs or longer make up satellite DNA. These satellite DNA strands are located close to the centromeres of chromosomes. Minisatellites are groups of tandem repetitions, each between 14 and 500 base pairs in length, whose combined length is often only a few thousand base pairs. Microsatellites are the smallest group; their repeat units range in size from 1 to 13 base pairs, and their entire array length is often less than a few hundred base pairs. Because they vary in length between individuals, minisatellites and microsatellites are of particular importance in human genetics because they are very valuable for gene mapping. In the human genome, a minisatellite or microsatellite can be found on average once every 2 kb; in total, they make up around 3% of the genome. Each human develops from a single-cell zygote (to a complex organism with roughly 100 trillion (10^{14}) distinct cells over the course of life. Few cells survive a lifetime, therefore new ones must be produced to replace those that die. The creation of new cells is necessary for both the replacement and the development processes. A cell must duplicate its internal components, including its DNA, prior to dividing; this takes place during interphase. The term cell cycle refers to the alternating phases of mitosis and interphase. The S (synthesis) phase is when DNA replication takes place. Some DNA repair occurs as the cell gets ready for mitosis during the G2 phase, which occurs between the S phase and the next mitosis. The cell has two copies of each of the 46 chromosomes by the time it reaches the G2 phase. Sister chromatids are the name given to these identical chromosomes. Sister chromatid exchange is the term for the material exchange that takes place between sister chromatids during interphase.

From one type of cell to another, the length of the cell cycle varies greatly. The cycle can be finished in less than 10 hours in cells that divide quickly, such as epithelial tissue. Some cells, like those in the liver, might only divide once every few years or so. In adults, some cell types, including neurons and skeletal muscle cells, largely lose their capacity for cell division and replication. Even while the length of each stage of the cell cycle varies to some extent, variations in the length of the G1 phase account for the most of these variations. It is common to refer to cells as being in the G0 stage when they have stopped dividing for a while. In response to crucial internal and external signals, cells divide. For instance, DNA replication must be timed and finished as well as the cell must have grown to the proper size before the cell begins mitosis. Complex molecular interactions mediate this control, requiring the cell to respond to external stimuli that demand increased or decreased rates of cell division. The cyclin-dependent kinases (CDKs), a family of kinases that phosphorylates other regulatory proteins at crucial points in the cell cycle, are among the most important of the molecules involved. The various cyclins, which are proteins that are generated at specific cell-cycle stages and are later eliminated when CDK action is no longer required, must form complexes with the CDKs in order for them to perform this function. Due to their crucial function in the cell cycle and the possibility that a malfunction could result in cancer, cyclins and CDKs, as well as proteins that interact with them, are the focus of research. Even while mitosis normally only lasts one to two hours, it involves a number of crucial and intricate processes. There are multiple phases of mitosis. The chromosomes become clearly visible under a light microscope during prophase, the first mitotic stage. Each chromosome has two sister chromatids that are joined at a location known as the centromere. During this phase, the nuclear membrane that encloses the nucleus disappears. Spindle fibers start to develop from two centrioles on the cell's opposite sides. The two sister chromatids are finally pulled apart by the spindle fibers, which grow connected to the centromeres of each chromosome. During metaphase, the following stage of mitosis, the chromosomes are in their densely packed condition. They are easiest to see under a microscope at this stage because they are highly condensed. This is why metaphase chromosomes are typically used in the clinical diagnosis of chromosome abnormalities.

6.9 SPINDLE FIBERS

The spindle fibers start to tighten and pull the chromosome centromeres, which are now positioned in the center of the spindle, during metaphase. The centromere of each chromosome separates during anaphase, the mitotic stage, allowing the sister chromatids to separate. The spindle fibers then drag the chromatids in opposite directions along the cell's axis, centromere first. The cell possesses 92 distinct chromosomes at the conclusion of anaphase, with half of them located near one side and the other near the other. New nuclear membranes are formed around each of the two sets of 46 chromosomes during telophase, the last phase of mitosis. Moreover, the chromosomes start to decondense and the spindle fibers disappear. Cytokinesis, which typically follows nuclear division, divides the cytoplasm into two roughly equal portions. Two diploid daughter cells that are both identical to the original cell have been created following the completion of telophase. The chromosomes of an egg cell and a sperm cell are joined to produce a zygote, which is a single cell. Given that humans are diploid organisms, a method to decrease the number of chromosomes in gametes to the haploid state must exist. Otherwise, the zygote would have 92 chromosomes instead of the typical 46. Meiosis is the main process by which haploid gametes are created from diploid progenitor cells (Zhou et al., 2016)although faithful replication of meiosis has proven to be a substantial obstacle to deriving haploid gamete cells in culture. Here we report complete in vitro meiosis from embryonic stem cell (ESC. During meiosis, two cell divisions take place. The stages of each meiotic division have the same nomenclature as those of mitosis, although some of the processes involved in these stages are very different. A diploid cell divides into two haploid cells during meiosis I, which is also known as the reduction division stage. The oogonia in females and the spermatogonia in males are these diploid cells. Following meiosis I, each haploid cell is reproduced via a second meiosis known as the equational division.

Interphase I is the first phase of the meiotic cell cycle, which is when crucial events like chromosomal DNA replication occur. Prophase I, the second phase of meiosis I, is very intricate and contains many of the crucial processes that set meiosis separately from mitosis. As the chromatin strands coil and compress and start to appear as chromosomes, prophase I starts. The homologous chromosomes pair up, side by side, lying together in perfect alignment during the process of synapsis. This homologous chromosome pairing is a crucial step in the cell cycle that does not take place during mitosis. The chromatids of the two chromosomes entangle as prophase I progresses. The number of chromosomes in each interwoven homologous pair is either bivalent or tetrad. Each chiasma marks the location of a genetic material exchange between homologous chromosomes. Chromosomes created by this process, known as crossing over, are composites of the original chromosome components (Stern, 1936). This chromosomal rearrangement is crucial because it dramatically expands the gene combinations that can exist in each gamete and, in turn, the combinations of human features that can exist. This phenomenon is crucial for determining the arrangement of genes throughout chromosomes. At the conclusion of prophase I, the nuclear membrane separates, the bivalents start to move in the direction of the equatorial plane, and a spindle apparatus starts to form in the cytoplasm. The following phase is metaphase I. The completion of spindle formation and alignment of the bivalents, which are still linked at the chiasmata, in the equatorial plane characterize this stage, just like in mitotic metaphase. Each bivalent's two centromeres are now located on opposite sides of the equatorial plane.

The homologous chromosomes are drawn toward the poles by the spindle fibers as the chiasmata disappear during anaphase I. This characteristic is in contrast to the comparable mitotic phase, the centromeres do not divide and replicate, resulting in just half of the initial number of chromosomes migrating toward each pole. Thus, one chromosome from each pair of autosomes and one sex chromosome makes up the chromosomes traveling toward each pole. Each of the two daughter cells has two sister chromatids and the same number of chromosomes as the mother cell. During this period, cytokinesis also takes place in humans. In the gametes produced by men, the cytoplasm is roughly split between the two daughter cells. One daughter cell, which will later develop into the egg, receives almost all of the cytoplasm when they are created in females. The other daughter cell eventually degenerates into a polar body, a tiny, inactive cell. Interphase II then marks the start of the equational division, meiosis II. Contrary to interphase I, there is no DNA replication during interphase II, which is a crucial characteristic. The following phase, prophase II, resembles mitotic prophase rather closely, with the exception that the cell nucleus only contains haploid chromosomes. The nuclear membrane disappears, the chromosomes thicken as they coil, and fresh spindle fibers are created during prophase II. Metaphase II comes next, when the chromosomes are aligned at the equatorial plane by the spindle fibers. The next step is anaphase II. In that the centromeres separate and each conducts a single chromatid toward a cell pole, this stage is similar to mitotic anaphase. The chromatids have now separated; however, the newly separated sister chromatids might not be identical due to chiasma development and crossing over.

Like telophase I, telophase II starts when the chromosomes reach the poles. They start to uncoil there. Each set of chromosomes is surrounded by fresh nuclear membranes, and cytokinesis takes place. Thus, four functional daughter cells with an equal amount of cytoplasm are the final product of meiosis. Again, uneven cytoplasmic division results in the formation of the egg cell and another polar body in female gametes. When the second stage of meiosis is complete, three polar bodies may exist because the polar body created during meiosis I occasionally undergo a second division. Errors that take place during meiosis are the primary cause of most chromosomal abnormalities. It is possible to produce gametes that have chromosomes missing, extra, or with different structures. Furthermore, early embryonic mitotic errors can impact enough cells in the body to result in a clinically severe disease. There is a lot of genetic diversity in humans. Characteristics like height, blood pressure, and skin tone reflect this. Disease states like cystic fibrosis and type 1 neurofibromatosis fall within the category of genetic diversity. Medical genetics is concerned with this element of genetic variation. Somatic cells and germ cells can both be affected by mutations.

Somatic cell mutations are of great concern since they can result in cancer. A DNA sequence can vary between individuals as a result of mutations. The various sequences are known as alleles. The term locus refers to a gene's chromosomal position. For instance, one could state that they have a certain allele at the chromosome 11 -globin locus. A person is said to be a homozygote if he/she shares the same allele on both members of a chromosomal pair. He or she is a heterozygote if the DNA sequences of the alleles are different. The genotype refers to the set of alleles that are present at a specific location.

The term mutation has frequently been used in human genetics to refer to DNA sequence alterations that result in genetic diseases and are therefore quite uncommon, with a population frequency of less than 1% (Chuzhanova, Anassis, Ball, Krawczak, & Cooper, 2003)a complex lesion that appears to represent a combination of micro-deletion and micro-insertion. In the absence of meta-analytical studies of indels, the mutational mechanisms underlying indel formation remain unclear. Data from the Human Gene Mutation Database (HGMD. The term polymorphism is commonly used to characterize DNA sequence variations that are more prevalent in populations. Polymorphic loci are those that have several alleles. However, these days, polymorphisms are also frequently used to refer to alleles with a frequency lower than 1%. Some mutations involve changes in the number of chromosomes within a cell. Base pair substitution, which involves switching out one base pair for another, is an important type of single-gene mutation. The amino acid sequence may vary as a result of this. However, many of these mutations do not alter the sequence of amino acids and hence often have no impact due to the redundancy of the genetic code. Silent substitutions are the name given to such alterations. There are two main types of base-pair substitutions that affect amino acids: missense mutations, which result in a change to a single amino acid, and nonsense mutations, which result in one of the messenger RNA's stop codons (UAA, UAG, or UGA). Because stop codons stop the translation of the mRNA, nonsense mutations cause the polypeptide chain to end prematurely. On the other hand, a polypeptide that is excessively lengthy can be created if a stop codon is changed so that it encodes an amino acid. Different mutations can change how transcription or translation are regulated. A promoter mutation can diminish the affinity of RNA polymerase for a promoter site, which frequently leads to decreased mRNA synthesis and hence decreased protein synthesis. Similar outcomes can result from altering enhancer sequences or transcription factor genes. As mature mRNA is created from the primary mRNA transcript, mutations can potentially interfere with the splicing of introns. Splice-site mutations, or those that happen at intron-exon borders, change the splicing signal required for accurate intron excision. Splice-site mutations can occur at the AG sequence, which defines the 3′ splice site, or at the GT sequence, which defines the 5′ splice site. Moreover, they may occur in sequences that are close to the donor and acceptor sites. When such alterations take place, the excision is frequently done at a splice point inside the following exon. These splice sites are typically underused and concealed within the exon, and their DNA sequences differ slightly from those of typical splice sites. Thus, they are known as cryptic splice sites.

6.10 SICKLE CELL ANEMIA

Sickle cell disease, which is caused by an anomaly in the structure of hemoglobin, affects 1 in 400(Russo et al., 2019). Typically, a single missense mutation that results in the replacement of valine for glutamic acid at position 6 of the beta-globin polypeptide chain causes sickle cell disease. In homozygotes, this amino acid alteration causes aggregation of hemoglobin molecules to form, which gives erythrocytes their distinctive sickle shape when there is low oxygen tension. Normal erythrocytes can fit through capillaries, but sickled erythrocytes are rigid and are unable to do so. The aberrant erythrocytes also have a propensity to adhere to the vascular endothelium, which is the innermost layer of blood vessels. Localized hypoxemia (lack of oxygen), and infarctions of many tissues, including bone, spleen, kidneys, brain, and lungs, are all caused by the resulting vascular blockage. Anemia is caused by the premature destruction of sickled erythrocytes, which lowers the level of hemoglobin and the number of circulating erythrocytes. Splenomegaly causes the spleen to grow, but infarctions eventually kill the organ, resulting in partial immune system function loss. This increases the risk of the recurrent and occasionally fatal bacterial infections that sickle cell disease patients frequently experience. Before the age of 20, about 10% of sickle cell disease patients have a stroke. The mutations that cause thalassemia diminish the amount of globin, as opposed to sickle cell disease, in which a mutation modifies the structure of the hemoglobin molecule. The deletions of the beta-globin genes account for the majority of cases of thalassemia. Three of the genes are lost or altered, resulting in HbH disease, which causes moderately severe anemia and splenomegaly. Hypoxemia in the fetus and hydrops fetalis are brought on by loss of all four genes, a syndrome that is most common in Southeast

Asians. Stillbirth or neonatal mortality are frequently caused by severe hydrops fetalis.

Heterozygotes who have a mutation in the globin gene on one copy of chromosome 11 are considered to have thalassemia minor, a disorder with little to no anemia and usually no need for clinical care. A person with a globin mutation in both copies of the chromosome either develops thalassemia major (Cooley's anemia) or a less severe form called thalassemia intermedia. Globin levels might be completely missing (0-thalassemia) or reduced to between 10% and 30% of normal levels. The disease phenotype of 0-thalassemia is often more severe, however, patients with 0-thalassemia who also have globin mutations that decrease the number of globin chains are less severely impacted since disease features are brought on by an excess of globin chains. Since globin is not generated until after birth, thalassemia symptoms do not manifest clinically until between 2 and 6 months of age. Severe anemia sets in for some people. If the disease is not treated, significant growth retardation may take place. The skeletal alterations brought on by the anemia include a protruding upper jaw, protruding cheekbones, and weakening of the long bones. Patients with untreated thalassemia major frequently die in their first ten years of life, and splenomegaly and infections are frequent. A promoter, two enhancers, and an area upstream known as the locus control region (LCR) govern the transcription of globin. In most cases, mutations in these regulatory areas lead to decreased mRNA production and decreased, but not entirely absent, globin. Splice-site mutations of various varieties have also been noted.

6.11 POINT MUTATIONS

Point mutations at the donor or acceptor sites entirely impair normal splicing, resulting in 0-thalassemia (Rogan, Faux, & Schneider, 1998)bits. Usually, mutations in the nearby consensus sequences cause thalassemia. The cryptic splice sites located in the introns or exons of the globin gene can also become mutated, making them accessible to the splicing mechanism. The result of this competition between the extra splice sites and the regular splice sites is the production of both normal and aberrant globin chains. The usual outcome is thalassemia. There have been claims to be hundreds of different globin mutations. Due to this, the majority of thalassemia patients are not homozygotes in the strict sense; instead, they are considered compound heterozygotes since they typically carry a separate globin mutation on each copy of chromosome 11. Each of the two globin genes is mutated, resulting in a disease state, even if the mutations vary. Compound heterozygotes are often loosely referred to as homozygotes. Blood transfusions and chelating medications that remove extra iron introduced by the transfusions are sometimes used to treat sickle cell disease or thalassemia major patients. Antibiotics and the anti-pneumococcal vaccine are given as preventative measures to sickle cell disease patients to help avoid bacterial infections, and analgesics are given to relieve discomfort when problems occur. The death rate from this treatment is still rather high, and it is frequently impossible to find a donor who is well matched. Moreover, gene therapy is a strong option for treating thalassemia. It is known that several different substances can result in induced mutations.

In contrast to spontaneous mutations, which occur naturally in cells, such as during DNA replication, these mutations are related to established environmental factors (Wogan, Hecht, Felton, Conney, & Loeb, 2004)air or water, and also endogenously when they are products of metabolism or pathophysiologic states such as inflammation. It has been estimated that exposure to environmental chemical carcinogens may contribute significantly to the causation of a sizable fraction, perhaps a majority, of human cancers, when exposures are related to "life-style" factors such as diet, tobacco use, etc. This chapter summarizes several aspects of environmental chemical carcinogenesis that have been extensively studied and illustrates the power of mechanistic investigation combined with molecular epidemiologic approaches in establishing causative linkages between environmental exposures and increased cancer risks. A causative relationship between exposure to aflatoxin, a strongly carcinogenic mold-produced contaminant of dietary staples in Asia and Africa, and elevated risk for primary liver cancer has been demonstrated through the application of well-validated biomarkers in molecular epidemiology. These studies have also identified a striking synergistic interaction between aflatoxin and hepatitis B virus infection in elevating liver cancer risk. Use of tobacco products provides a clear example of cancer causation by a life-style factor involving carcinogen exposure. Tobacco carcinogens and their DNA adducts are central to cancer induction by tobacco products, and the contribution of specific tobacco carcinogens (e.g. PAH and NNK. Mutagens are all substances that result in induced mutations. Radiation is an important class of mutagens. Ionizing radiation, which includes x-rays can cause atoms to lose their electrons, resulting in the formation of electrically charged ions. These ions can stimulate chemical processes that alter DNA bases when they are present inside or close to the DNA molecule. Moreover, double-stranded DNA can be broken by ionizing radiation. It is possible for this type of radiation to enter germ cells as well as other bodily or somatic cells. Although it does not produce charged ions, non-ionizing radiation has the ability to transfer electrons from an atom's inner to outer orbits. The atom develops a chemical instability. Non-Ionizing radiation includes ultraviolet (UV) radiation, which the sun naturally produces. Adjacent pyrimidine bases create covalent bonds with one another due to UV light. When these pyrimidine dimers try to join with purines during DNA replication, they are unable to do so properly, which leads to a base-pair substitution. Skin cancer can be caused by UV radiation, which is absorbed by the skin and does not reach the germ cells. It is challenging to directly assess a mutation in people because it is an infrequent occurrence. It is also challenging to determine how radiation exposure and mutation interact. A normal range of lifetime exposure to ionizing radiation for a person in a developed nation is between 6 and 7 rem.

Some human groups have been exposed to even higher radiation levels. The survivors of the atomic bomb explosions that took place in Hiroshima and Nagasaki at the end of World War II are the group that has been the subject of the most extensive research. Radiation claimed the lives of many people who received high radiation exposures. Many of the survivors gave birth to children, while others survived. An extensive group of Japanese and American scientists examined some of the medical and genetic records of survivors to learn more about the effects of radiation exposure in this community. Many of them, perhaps as a result of radiation exposure, developed malignancies and chromosome abnormalities in their somatic cells. The scientists compared the offspring of individuals who had significant radiation exposure with the offspring of those who did not in order to gauge the impact of radiation exposure on the patients' germ cells. The children of those who were exposed to radiation and the children of those who were not did not differ statistically significantly. Moreover, utilizing minisatellite polymorphisms and protein electrophoresis, a method that finds mutations that result in amino acid alterations, direct genetic investigations of mutations have been conducted. To ascertain whether germline mutations had occurred at various loci, parents and offspring were compared. Statistics showed that the number of mutations found in the exposed and unexposed groups was equal. This illustrates how somatic mutations have an impact. However, it is still uncertain whether there is increased evidence for protein-coding DNA germline mutations.

6.12 CANCER

Leukemia and thyroid cancer rates have gone up in some people as a result of nuclear testing (Moysich et al., 2002). Some houses contain dangerously high levels of radon, a radioactive gas that causes lung cancer when it breaks down naturally occurring uranium. Any unwanted radiation exposure should be avoided, especially when it comes to the gonads or developing babies.. Many substances can cause mutations as well, often as a result of their chemical resemblance to DNA bases. These base analogs, like 5-bromouracil, can be used in place of actual DNA bases during replication because of this resemblance. Since the analog does not perfectly match the base it replaces, pairing errors may occur during future replications. Other chemical mutagens can physically insert themselves between already-existing bases, twisting the DNA helix and resulting in frameshift mutations. Examples include acridine dyes. Other mutagens can directly change DNA bases, leading to errors in DNA replication. It is known that hundreds of substances can cause mutations in lab animals. These include sodium nitrite, formaldehyde, vinyl chloride, alkylating agents, and saccharin. These substances vary greatly in their mutagenic potency. Humans have long been fascinated by the individual distinctions that can be observed. Long ago, attention was drawn to apparent variations like skin tone or body type. It wasn't until the 20th century that it was feasible to study genetic variety, which is the result of cumulative mutations through time. In order to map genes to precise chromosomal sites, a crucial step in figuring out how genes operate, it is vital to evaluate and analyze this variation in populations and families. Evaluation of genetic variation is crucial in forensics since it forms the basis for genetic diagnosis.

Antigens found on the surfaces of erythrocytes have been used to define a number of blood group systems. Some factors are taken into account while deciding whether a person can receive blood from a certain donor. Different genes or sets of genes control the blood group systems. The varied DNA sequences in these genes produce the antigens that can be produced inside a system. Some of these blood type combinations can cause maternal-fetal incompatibility, which can occasionally have consequences for the fetus. Even while human blood transfusions had been attempted as early as 1818, they frequently failed. Following transfusion, some recipients experienced a severe, occasionally fatal hemolytic response. Karl Landsteiner in 1900 made the discovery that the ABO antigens on the surface of erythrocytes were responsible for this reaction. A and B are the names of the two main antigens that make up the ABO system. Each person contains antibodies that respond to any antigens that aren't on the surface of their own red blood cells. For instance, if type B blood is transfused into someone who has anti-B antibodies, a strong antibody reaction will result. By combining a small sample of blood with solutions containing various antibodies and observing which combinations result in the observable clumping that is typical of an antibody-antigen interaction, it is simple to detect an individual's ABO blood type in a lab setting. Early research, for instance, revealed that the B antigen is very prevalent in Asian people and the A antigen is relatively common in populations of Western Europeans. Both of these antigens are uncommon in aboriginal populations of South America, where blood type O makes up the majority. Similar to the ABO system, the Rh system is based on the antigens that are found on the surfaces of erythrocytes. Rh alleles are another helpful tool for determining genetic variation because they vary greatly among individuals and populations.

Protein electrophoresis, which was initially devised in the 1930s and was extensively used on humans in the 1950s and 1960s, significantly increased the number of identifiable polymorphic loci. This method takes advantage of the fact that a single amino acid or AA change in a protein resulting from a mutation in the associated DNA sequence can result in a small change in the protein's electrical charge. Due to the fact that valine only has one carboxyl group but glutamic acid has two, the electrical charge changes when valine is substituted for glutamic acid in the globin chain. Using electrophoresis, one may identify if a person has normal hemoglobin (HbA) or hemoglobin with the sickle cell disease mutation (HbS). An electrically charged gel made of starch, agarose, or polyacrylamide is used to hold the hemoglobin. The HbA and HbS forms migrate through the gel at various rates due to the minute change in charge brought on by the differing amino acid composition. After allowing the protein molecules to migrate for several hours, they are dyed with chemical solutions to make it possible to see where they are. It is possible to tell from the resulting pattern whether a person is a HbA homozygote, a HbS homozygote, or a heterozygote with HbA on one copy of the chromosome and HbS on the other. Many human proteins were subjected to protein electrophoresis in order to identify variations in amino acids. This method cannot identify silent substitutions, which do not change the amino acids. Moreover, some amino acid modifications leave the protein molecule's electrical charge unaffected. For these reasons, only roughly one-third of the changes that occur in coding DNA are picked up by protein electrophoresis. In addition, protein electrophoresis typically misses single-base alterations in non-coding DNA.

These methods have only been able to detect a very small portion of human DNA variation since there are only a few hundred blood types and protein electrophoretic polymorphisms. However, evaluating this variance is important for both gene discovery and genetic diagnosis. Fortunately, molecular methods created since the 1980s have made it possible to find millions of new DNA polymorphisms. The tens of thousands of copies of human DNA required for Southern blotting and other experimental uses can be produced by cloning. The production of genetically modified therapeutic products, such as insulin, interferon, human growth hormone, clotting factor VIII, and tissue plasminogen activator is also currently carried out using this method. Gene products that have been genetically altered are quickly replacing natural ones as a more affordable, pure, and effective option. Southern blotting has a variety of applications. It can, for instance, identify insertions or deletions in DNA sequences that enlarge or reduce particular DNA fragments. This method can be employed as a low-cost and effective diagnostic tool if a disease-causing mutation modifies a specific restriction site, as in the case of sickle cell disease. Restriction fragment length polymorphisms (RFLPs), which are present throughout the human genome as a result of typical DNA sequence variation, were analyzed by Southern blotting. Many disease-causing genes, such as those that cause cystic fibrosis, Huntington's disease, and type 1 neurofibromatosis, were localized utilizing these sequence variants (Botstein & Risch, 2003).

The number of repeats in a given region, which differs greatly from person to person and is used to evaluate genetic variation, can contain as few as two or three repeats or as many as 20 or more. Therefore, these polymorphisms can show a substantial amount of genetic variation. Microsatellite polymorphisms are known as short tandem repeats (STR), while minisatellite polymorphisms are known as variable number of tandem repeats (VNTR). Single nucleotide polymorphisms, which are the most prevalent type of polymorphism in the human genome, are variations at specific chromosomal locations. RFLPs are a subset of the larger set of SNPs and are typically brought on by single-base changes that only exist at restriction sites. Even though the majority of these polymorphisms are benign, when they appear in functional DNA sequences, they can result in inherited diseases. Microarray and direct sequencing techniques are increasingly being used to find them. Therefore, genetic variation may be utilized to identify people, much like a traditional fingerprint. Genetic diversity has potential in forensic applications because DNA may be detected in any tissue sample, including blood, semen, and hair. With their many alleles,

STRs are excellent for creating a highly precise DNA profile. A DNA profile operates on a simple principle. The probability that any other person in the population has the same allele at each analyzed location becomes incredibly small if we examine enough polymorphisms in a particular individual. Even a little sample that is several years old can produce enough DNA for laboratory analysis due to the extraordinary sensitivity of the PCR method. The discovered alleles are then contrasted with a suspect's alleles. The small size of the DNA molecule prevents direct visualization of DNA variation. All techniques for measuring DNA variation include indirect measurement, such as when Southern blotting is used to bind tagged probes to certain DNA sections. Most techniques for displaying DNA variation call for the indirect tagging of DNA. Multiple copies must be created in order to observe the labels. For instance, the labeled probes used in Southern blotting can be replicated thousands of times using microorganisms. However, this procedure is laborious, frequently taking several days or longer, and it frequently needs a sizable amount of DNA from the individual. The discovery of an alternate method, the polymerase chain reaction (PCR), in the middle of the 1980s greatly improved the efficiency of genetic variant identification at the DNA level.

Chapter

7

EPIGENETICS

CONTENTS

7.1 INTRODUCTION

Conrad Waddington coined the word epigenetics (Figure 7.1) in 1942, and it was later described as the causal interactions between genes and their products that permit phenotypic expression (Deichmann, 2016).

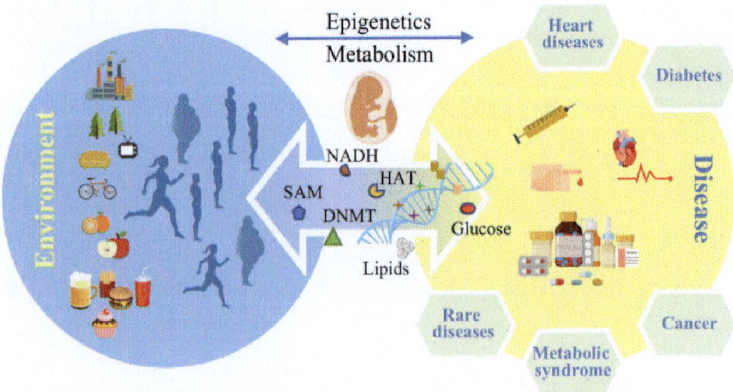

Figure 7.1. Epigenetics and health. (Source: Tzika E, Dreker T and Imhof A (2018) Epigenetics and Metabolism in Health and Disease. Front. Genet. 9:361. doi: 10.3389/fgene.2018.00361, Creative Commons License).

Although there are many different definitions of this term nowadays, a generally accepted meaning is that epigenetics refers to all heritable changes in phenotype brought on by mechanisms unrelated to the core DNA sequence. For a very long time, it was believed that cellular divisions were the only time that epigenetic information could be passed down. However, it is now known that organisms can pass on epigenetic processes from one generation to the next. The DNA methyltransferases (DNMTs) are responsible for catalyzing this crucial transfer step. The three primary DNMTs are DNMT1, 3A, and 3B, and DNMT1 catalyzes the methylation that takes place as the DNA is copied during each cellular replication. The other two main DNMTs, 3A and 3B, are more well-known for their comparatively increased de novo methylation activity, which occurs when new 5-methyl cytosines are added to the genome at locations that have not yet been methylated. Although there are some notable exceptions to this rule, in general, the more methylated a gene regulatory region is, the more probable it is that the gene activity will become down-regulated and vice versa. Recent research has emphasized the critical functions of UHRF1 and DNMT3L, which are needed for maintenance and de novo methylation, respectively, as well as the possibility of combining 5-hydroxymethylcytosine with 5-methylcytosine to express the effects of DNA methylation on the genome.

7.2 MECHANISMS

Another essential epigenetic mechanism that affects many other biological functions in addition to gene expression is chromatin modifications. Histones undergo site-specific post-translational changes, such as acetylation and methylation, which affect the functions of other proteins involved in gene regulation. Key histones that make up the chromatin are acetylated by histone acetyltransferases (HATs), and they are deacetylated by histone deacetylases (HDACs) (Dokmanovic & Marks, 2005)inhibitors represent a new class of targeted anti-cancer agents. Several of these compounds are in clinical trials with significant activity against a spectrum of both hematologic and solid tumors at doses that are well tolerated by the patients. The HDAC inhibitors are a structurally diverse group of molecules that can induce growth arrest, differentiation, apoptosis, and autophagocytic cell death of cancer cells. While the base sequence of DNA provides the genetic code for proteins, the expression of genes is regulated, in large part, by the structure of the chromatin proteins around which the DNA is wrapped (epigenetic gene regulation. These variations can occur at a variety of locations in the histones but are most frequently seen in the amino-terminal regions of these proteins. Histone acetylation and gene activity are typically correlated, and vice versa. For instance, non-coding RNA, including both its short and long forms, frequently shares protein and RNA components with the RNA interference (RNAi) pathway and may also have an impact on epigenetics including DNA methylation (Kaikkonen, Lam, & Glass, 2011). These impacts seem to affect a wide variety of organisms, from protists to humans. Prions are interesting because they have epigenetic effects that are independent of DNA and chromatin

(Harvey, Chakravarty, Futia, & Jarosz, 2020). Structural heredity is crucial for epigenetic expression, where different states of regulatory networks or macromolecular complexes can affect phenotypic expression without affecting DNA sequences. The ability of the prion proteins to change their structure through an autocatalytic process can affect epigenetic expression and cause disease in humans. The expression of a gene can also be influenced by its location on a particular chromosome.

7.3 ADVANCEMENTS

Furthermore, there have been significant advancements in high-resolution methylome analysis. New methods for analyzing genome-wide aspects of epigenetics, based on the use of methylation-sensitive restriction enzymes, sodium bisulfite conversion, and affinity capture with antibodies or proteins that select methylated DNA sequences, have been made possible by microarray platforms and high-throughput sequencing. Methods such as targeted and whole genome bisulfite sequencing, methylated DNA immunoprecipitation (MeDIP), methylation-sensitive restriction fingerprinting, and digital karyotyping specific to methylation were formulated. It is possible to produce a three-dimensional image of the genome inside the nucleus by analyzing the spatial closeness of distant functional genomic locations. The methods to interpret data from ChIP-on-chip and ChIP-seq, for example, are becoming more crucial due to the abundance of information obtained from epigenomic technologies. There is no doubt that improvements in the methods for analyzing epigenetic data have been and will continue to be key drivers of epigenetic development. Epigenetic processes are prevalent, and much of our current understanding of epigenetics comes from model systems that can be both common and exceptional. The simplicity of working with eukaryotic microorganisms has aided in the discovery of the molecular mechanisms underlying fundamental epigenetic events. In these situations, epigenetics may be crucial in understanding the significance of gene silencing mechanisms in evolution as well as protecting the genome from intrusive DNA elements.

7.4 DROSOPHILA

Drosophila helps in understanding the evolutionary implications of epigenetics (Hu & Barrett, 2017). The mouse model is the most effective model system for epigenetics. Randy Jirtle and colleagues analyzed a wide range of mice models that are crucial for epigenetic processes like imprinting and trans-generational epigenetics and have the potential to provide light on human diseases like diabetes, cancer, and neurological disorders. In epigenetics, plant models are crucial because of their plasticity and capacity to silence transposable elements (Grant-Downton & Dickinson, 2006). RNAi silencing in plants has been at the forefront of epigenetics. Epigenetics and alterations in an organism's metabolism are tightly intertwined, making it impossible to completely comprehend each process in isolation. The universal methyl donor S-adenosylmethionine (SAM) powers numerous epigenetic processes, and its role in epigenetic mechanisms is crucial (Mahmood et al., 2017). Chromatin, a key modulator of epigenetic events, can be affected by metabolic activities. It is now known that a variety of environmental factors and metabolic substances can control the enzymes that animals use to change their histones. Therefore, the two main epigenetic mediators, DNA methylation and chromatin remodeling, are influenced by metabolic processes, and it is anticipated that this technology will continue to grow exponentially. Environmental signals are used by stem cells in part, and epigenetic processes like DNA methylation, histone modification, and microRNA (miRNA) play crucial roles in how stem cells react to environmental factors. Stem cells are essential to regenerative medicine, and skeletal muscle regeneration entails significant epigenetic modifications that control the expression of certain genes in muscle progenitors through chromatin as well as microRNA epigenetic modifications (Loscalzo & Handy, 2014).

Similar to genomic imprinting, epigenetics has been recognized to play a role in this process for a long time, although research in this field of epigenetics is still advancing quickly. Genomic imprinting can occur in specific tissues or at crucial developmental phases, and it is not just found in mammals but also in plants and animals through comparable processes. The epigenetics of memory processing has lately undergone development. Recent interesting discoveries have demonstrated that changes in adult brain function and behavior depending on life events require gene regulation through epigenetic processes. In addition, novel medications that affect epigenetic pathways may one day be used to treat or lessen cognitive impairment. Trans-generational inheritance is also a sort of memory-based in part on epigenetics, as it can significantly affect adult health and disease risk depending on early life events that have an impact on epigenetic markers. The aging process is a type of epigenetic memory and experience in that it alters our genes epigenetically throughout our lives and as a result of our parents. This can have an impact on human longevity and our risk of developing a variety of age-related diseases, many of which are also epigenetically based. Therefore, it is clear that epigenetics affects a variety of distinct activities, and it is very possible that many more will be identified in the future. Even now, with the constantly shifting environment that can influence gene expression through epigenetic processes, the evolutionary impact of epigenetics is high. For instance, fast changes in diet, modern lifestyle, and environmental contamination have an impact on not just the epigenome of humans but also the evolution of many more primitive species, which in turn have a significant impact on the environment. Dietary components vary greatly not only between people but also among communities of humans and different animals. For instance, viruses and bacteria have an impact on how the genome's epigenetic expression is expressed, and these changes can result in diseases like cancer in humans. Walter Doerfler and colleagues discuss the specifics of how adenovirus type 12 (Ad12) altered the hamster genome (Doerfler, 2007). The new discipline of pharmacoepigenetics has been created as a result of how drugs also alter the epigenome. It is evident that various populations react to medications in different ways, and a large portion of this diversity may be explained by epigenetic variables. The implication of epigenetic modifications in the etiology, progression, and diagnosis of human diseases has sparked interest in epigenetics among the medical community. DNA methylation, chromatin modifications, and RNA-dependent regulation have all been demonstrated to influence the occurrence and severity of cancer, which has long been linked to epigenetic changes. Epigenetic abnormalities have been linked to a variety of immunological disorders, including systemic lupus erythematosus (SLE) and rheumatoid arthritis, as well as autoimmune disorders, including multiple sclerosis. Epigenetic processes have also been connected to brain disorders. Epigenetic changes have been linked to a variety of diseases, including Rett syndrome, Alzheimer's disease, Huntington's disease, and even autism, to mention a few (Qureshi & Mehler, 2018). Even the expression of schizophrenia and depression may have an epigenetic foundation. Epigenetic abnormalities may potentially be connected to system metabolic problems. For instance, hypertension, gestational diabetes, and obesity might affect the fetal chromatin and raise the risk of adult diseases later in life. It should not be surprising that errors in imprinting might cause a variety of human disorders as genomic imprinting is based on epigenetic mechanisms. Imprinting disorders caused by epigenetic abnormalities include the Prader-Willi syndrome, Angelman syndrome, silver Russell syndrome, and many more, including transitory neonatal diabetes mellitus. Because so many diseases are affected by epigenetic factors, improvements in their management will probably be dependent in part on new developments in the field of epigenetic therapy.

7.5 DRUGS

Histone modifying drugs have likely drawn the greatest interest in the clinics despite the fact that there are other epigenetic therapies in use (Taby & Issa, 2010). Histone deacetylase (HDAC) inhibitors are the most important of them. HDAC inhibitors come in a wide variety of forms, and it is likely that significant advancements will be made for HDAC inhibitors as well as many other medications that can normalize abnormalities in DNA methylation, histone modifications, and perhaps some of the many other epigenetic processes that have been identified. Histones are modified by a number of protein-modifying enzymes, especially at their flexible N-termini, including methyltransferases. When it comes to distributed retrotransposons and altered sex-specific de novo methylation of DNA sequences in germ cells, the phenotype of Dnmt3L knockout mice is identical to that of a Dnmt3a germ cell-specific conditional knockout. According to these findings, Dnmt3a and Dnmt3L are both necessary for the methylation of the majority of imprinted loci in germ cells. Dnmt3L promotes de novo methylation by both of these MTases and co-localizes and immunoprecipitates with Dnmt3a and Dnmt3b. The C-terminal domains of Dnmt3a and Dnmt3L contain the distinctive fold of Class I AdoMet-dependent MTases (Jia, Jurkowska, Zhang, Jeltsch, & Cheng, 2007). However, only Dnmt3a and not Dnmt3L contained S-adenosyl-L-homocysteine, a byproduct of the methylation reaction. This is consistent with Dnmt3a serving as the complex's catalytic component and Dnmt3L being dormant and unable to bind AdoMet. The total length of the Dnmt3a/Dnmt3L C-terminal complex is approximately 16 nm, longer than a core nucleosome's diameter of 11 nm. Two Dnmt3a monomers and two Dnmt3L monomers are present in this complex, generating a tetramer with two 3L-3a interfaces and one 3a-3a interface.

Dnmt3a and Dnmt3b have the shortest DNA binding domains of all known active DNA MTases (Chen, Tsujimoto, & Li, 2004). But when two active sites are brought together by dimerization at the 3a-3a contact, the DNA-binding surface is virtually doubled. The frequency of CpG sites in the differentially methylated regions of 12 maternally imprinted mouse genes has a similar pattern. These findings point to a rationale for the detection of both CpG spacing and nucleosome alteration in order to identify and methylate differentially methylated sites in imprinted genes. When embryonic stem cells were forced to differentiate, non-CG methylation disappeared, but it reappeared in induced pluripotent stem cells. Similar to this, it has been discovered that Arabidopsis thaliana DRM2, which is related to mammalian Dnmt3a, exhibits a 10-bp correlation of non-CpG DNA methylation. Histone modifications and DNA methylation are closely related to one another. DNA methylation is actually thought to be linked with histone methylation. Predominantly, the absence of H3K4 methylation and the presence of H3K9 methylation are associated with DNA methylation. Gene promoters may be shielded from de novo DNA methylation in somatic cells by histone H3 lysine 4 (H3K4) methylation. The H3K4 methylation patterns, which are not known to be preserved during chromosome replication, could be converted into heritable patterns of DNA methylation that mediate transcriptional silencing of the affected sequences by the mammalian de novo DNA methylation Dnmt3L-Dnmt3a machinery. *In vitro*, Dnmt3a is completely active on nucleosomal DNA. In addition to being necessary for genomic imprinting, Dnmt3a2 is a shorter variant of Dnmt3a that is predominate in embryonic stem cells and embryonal carcinoma cells and is also found in the testis, ovary, thymus, and spleen. The four core histones Dnmt3a2 and Dnmt3b were found to be the primary *in vivo* interaction partners of epitope-tagged Dnmt3L. When the amino tail of H3 and Dnmt3L were crystallized together, it was clear that the H3 amino tail had bound to the Cys-rich 3-Zn binding domain of Dnmt3L. Substituting specific residues in the binding site also rendered the H3-Dnmt3L association ineffective.

7.6 ENZYMES

Nucleotide modifying enzymes, such as DNA MTases, DNA repair enzymes, and RNA modification enzymes, frequently employ base flipping as a conserved method (Motorin & Helm, 2011). The DNA MTase domain and the SRA do not appear to share any structural or sequence similarities (or of DNA repair enzymes). However, the binding of the flipped base in a concave pocket, the use of two loops to simultaneously approach DNA from the major and minor grooves, and the phosphodiester backbone due to extensive protein-phosphate bonds surrounding the flipped nucleotide are analogous to the DNA MTases. Given that the SRA domain's recognition sequence consists of only two base pairs, this may be very significant. TET1 was overproduced in human cells, which resulted in the emergence of 5hmC. DNA 5mC simultaneously decreased, indicating that 5hmC is an oxidation byproduct of 5mC (Tan & Shi, 2012). 5hmC was found in ES cells, and RNAi-mediated reduction of TET1 lowered this percentage. The unexpected discovery of a 5mC oxidation pathway presents a number of issues, such as whether 5mC oxidation is a significant epigenetic change, either as a final result or as a step in active DNA demethylation, as suggested by the existence of 5hmC DNA excision repair glycosylase. Targeted deletion of the Dnmt2 gene revealed that Dnmt2 is not necessary for overall DNA methylation or maintenance in embryonic stem cells. While Dnmt2 protein expressed by bacteria failed to methylate DNA in vitro, Dnmt2 protein expressed by baculoviruses either showed no detectable activity or residual action towards DNA. A DNA methyltransferase-like catalytic mechanism is used by human Dnmt2 to methylate RNA cytosine-C5, which is not surprising given the high levels of sequence and structural conservation as well as the similar chemistry of DNA and RNA cytosine. It is impossible to create a chronological timeline of events or establish causal connections in gene silencing without an understanding of the interconnections and spatial links between their modular domains, as well as whether or not inter-domain interactions contribute to target specificity.

There are several conditions where gene or region-specific DNA methylation analysis is required, and there are many techniques available. It is critical to understand the specifics of each technique, such as the quantity of DNA needed, the flexibility with which CpG sites can be chosen for analysis, the degree of quantitative analysis possible, the level of technical difficulty, and the cost. For instance, if one wants to examine DNA methylation as a factor in gene silencing, one should examine a specific region that regulates gene expression, and a method that allows for flexible region selection should be utilized. A highly accurate procedure should be used if one wants to use it for diagnostic purposes. Several concepts can be used to assess DNA methylation and distinguish 5-methylcytosine (Cm) from cytosine (C). The first principle is dependent on restriction enzymes that are methylation-sensitive and that activity is influenced by the presence of a methyl group on a cytosine at a CpG site or sites within restriction sites. In contrast to how slowly it transforms methylated C, this procedure converts unmethylated C into uracil (U) rather quickly. DNA is digested by a methylation-sensitive restriction enzyme before being hybridized with a probe for a particular genomic region in Southern-blot hybridization for DNA methylation analysis. By keeping track of the band locations of the DNA fragments surrounding the restriction sites, it is possible to determine the methylation status of a restriction recognition site. Southern blot methodology is particularly helpful for analyzing repeated sequences because a single probe can examine a number of related sequences throughout the genome. However, this method uses a lot of high-quality DNA and only examines a small number of CpG sites that are contained within restriction recognition sites. Before bisulfite conversion-based approaches gained popularity, this technique was often employed, but it hasn't been used as much lately.

7.7 PCR

PCR is used to amplify bisulfite-converted DNA using primers positioned in genomic areas devoid of CpG sites (Leitão et al., 2018). A CpG site that has undergone methylation (C) and thymine (T) in the converted DNA reveals methylated and unmethylated C in the original DNA, respectively. Using this method, we may examine the methylation state of each and every CpG site located between the primers as well as the methylation patterns of many CpG sites within a single DNA molecule. This technique allows for the analysis of practically the DNA methylation. This method is labor-intensive and necessitates the sequencing of at least 10 clones from each sample, which could be a drawback. After bisulfite conversion, a restriction enzyme recognition site may develop. By performing PCR with primers specific to methylated or unmethylated sequences and checking for the presence or absence of a PCR product, this method probes the methylation statuses of multiple CpGs at primer locations. As long as enough of a PCR product is acquired for cloning, PCR cycles should be limited, even when ideal PCR conditions are used. In the absence of Taq polymerase activity, excessive PCR cycles result in the denaturation of the PCR product and the amplification of chimera products and even PCR products that weren't present in the template DNA. The difference in PCR efficiency between methylation and unmethylated DNA is also exaggerated by excessive PCR cycles. DNA methylation levels can be accurately, sensitively, and quantitatively assessed using real-time MSP. As in other investigations, a genomic region should be carefully chosen. In the same region, primers specific to methylation or unmethylated DNA should be created. A primer's 3' end should be at a polymorphic C/T site, and several CpG sites should be close to the 3' end. It can be challenging to create primers that are specific to unmethylated DNA, and using the bottom strand of the other DNA strand is typically useful.

The fully methylated and totally unmethylated DNA controls should be used to optimize the annealing temperature and magnesium concentration. There is sufficient amplification of fully methylated DNA and no amplification of entirely unmethylated DNA in the presence of methylation-specific primers. The amplification curve and melting curve can be used to estimate real-time MSP's ideal conditions. The optimal PCR conditions produce a single sharp peak on the melting curve. Standard DNA with known amounts of DNA molecules is required in order to quantify DNA methylation levels by real-time MSP. To start, a gel-filtration column can be used to remove any leftover primers and nucleotides from the PCR result. Second, a restriction enzyme linearizes the plasmid once the PCR product of MSP has been cloned into it. The number of DNA molecules in a solution can be computed because the molecular weight of the PCR product or the plasmid with the insert can be determined. Cloning a PCR product to create standard DNA provides the benefit of accuracy and the availability of a large quantity of standard DNA, but the process is labor-intensive. High sensitivity can be attained using MSP and real-time MSP, such as the ability to find one methylated DNA molecule out of 1,000. However, during bisulfite-mediated conversion, a substantial reduction in the number of DNA molecules that may act as PCR templates occur. Specifically, even though the weight of the DNA only slightly reduces, the quantity of template DNA molecules detected by quantitative PCR drops to 5 to 10% of the DNA before the treatment. Because of this, care must be taken in determining how many copies of template DNA are present in a PCR solution.

Regional DNA methylation analysis is used for diagnostic as well as fundamental research. To collect reliable data, it is necessary to choose an acceptable technique and carry out experiments under favorable conditions. The initial epigenetic alteration to be identified was DNA methylation. Both the biological effects of DNA methylation and the mechanisms regulating these modifications are well understood. In 1975, it was first suggested that DNA methylation played a functional role in cell differentiation, X chromosome inactivation, gene regulation, and epigenetic control. Currently, researchers are examining the relationships between DNA methylation and other epigenetic markers like histone alterations. Many laboratories examine and report abnormalities in DNA methylation patterns as a cause or result of disorders, such as cancer. Many strategies have been created to investigate DNA methylation profiles at the genome level. However, by coupling them with intensive cloning and sequencing effort or merging them with microarray platforms, genome-wide analytic tools have been established. The majority of the approaches were first utilized for detecting methylation alterations at the single gene level. Promoter or CpG island arrays are frequently employed on microarray platforms to examine crucial regulatory areas. Using arrays, the complete genome or selected chromosome regions can be studied. On the basis of their underlying principles, the majority of methylation analysis techniques can be divided into many distinct classes. On restriction endonucleases with changed sensitivity to methylation cytosine residues in the cleavage site, several techniques are based. In this manner, the methylation state of the cleavage sites influences the restriction endonuclease digestion pattern, which in turn reflects the methylation profiles of the specific chromosomal region. Other methods rely on proteins or antibodies that attach to DNA that has been methylated. High throughput techniques are now being developed for bisulfite sequencing methods, which are frequently needed for resolution at the single nucleotide level (Smith, Gu, Bock, Gnirke, & Meissner, 2009)reduced representation bisulfite sequencing (RRBS.

7.8 MSRF

One of the earliest techniques that allowed researchers to pinpoint many de novo methylated or unmethylated sequences occurring in various tissue samples was methylation-sensitive restriction fingerprinting (MSRF). For instance, tissue-derived DNA samples are first digested with MseI, a restriction enzyme that almost never cleaves within CpG islands. The information required for target gene identification and database searches is obtained following the cloning and sequencing of the amplified bands. The usage of MSRF is currently restricted due to the demand for gel electrophoresis. A genome-wide analytic technique called Methylation-Sensitive Representational Difference Analysis (MS-RDA) identifies DNA regions that differ in their levels of methylation between two samples. The fact that MS-RDA enriches the genome's unmethylated CpG-rich portion is one of its key features. Using a universal adaptor primer and a methylation-sensitive restriction enzyme, MS-RDA amplifies DNA fragments from genomic DNA. Differentially methylated DNA fragments between samples will appear in one amplicon but not the other. The two amplicons are used for genomic subtraction, and the resultant fragments are then cloned and sequenced. Differential methylation hybridization (DMH), which enables the simultaneous measurement of the methylation state of a significant number of CpG-island sites, is another illustration of restriction enzyme-based methods. High-density arrays are gridded with CpG islands containing DNA fragments, genomic DNA from the target tissues is digested with methylation-sensitive enzymes, and the digestion products are then employed as templates for PCR after linker ligation. When screening for hypermethylated sequences, such as in cancer tissues, the resultant fragments are employed as probes.

Methylation-sensitive restriction enzymes and PCR using flanking primers, followed by sequence analysis of microarray probing, are the foundation of Methylated CpG Island Amplification Coupled Microarray (MCAM) (Oakes, La Salle, Robaire, & Trasler, 2006)a few quantitative techniques have been developed by employing the use of real-time PCR, though they require the additional step of sodium bisulfite conversion. Here we evaluate a technique that involves the digestion of non-sodium bisulfite-treated genomic DNA using methylation-sensitive and methylation-dependent restriction enzymes followed by real-time PCR. The utility of this method is tested by analyzing seventeen genomic regions of known tissue-specific levels of DNA methylation including three imprinted genes. We find that this approach generates rapid, reproducible and accurate results (range= ±5%. MCAM is a potent method for simultaneously identifying genomic areas with differential methylation. The genomic DNA is first cut with a methylation-sensitive enzyme which results in blunt ends and helps to remove unmethylated sequences from the adaptor ligation process. A proper adaptor ligation is made possible by the second enzyme, XmaI, which recognizes the same sequence regardless of methylation state and produces sticky ends. A microarray platform is hybridized with the adaptor-ligated methylated fraction after it has been PCR-amplified, fluorescently tagged, and amplified. Using MCAM, a variety of disease-related DNA methylation alterations can be easily found in normal and disease-derived tissue. The methylation-dependent restriction enzyme McrBC, which exclusively cleaves methylated DNA, is another version of the current methylation microarray methods. The method works well for genome-scale DNA methylation analysis even though the precise cleavage point of this enzyme does not directly coincide with the methylation site. Sodium bisulfite deaminates cytosine to uracil, but 5-methylcytosine is resistant to bisulfite-induced deamination, which forms the basis of sodium bisulfite sequencing. DNA methylation pattern analysis using single base resolution is made possible by bisulfite sequencing. It is based on the sequencing of PCR products of bisulfite-treated DNA to profile DNA methylation at particular loci or along large segments of chromosomes. This method should ideally be used for full genome methylation profiling. Except for the few remaining methylated cytosines, sodium bisulfite treatment of DNA essentially divides the genome into three bases (A, G, and U/T). Many medium to high-throughput techniques have been developed to evaluate the methylation profiles of multiple genes at once after the conversion of genomic DNA with sodium bisulfite.

The bisulfite-pretreated genomic DNA is employed as a template for the PCR amplification of specific genes in the mass spectrometry-based MALDI-TOF MS. One of the PCR primers has a T7 RNA polymerase promoter linked to the 5' end by design. This promoter is used to convert the amplicon into single-stranded RNA *in vitro*, which is then digested with ribonuclease A at uracil residues. A medium-throughput method called MALDI-TOF MS delivers direct quantitative methylation data for many genes at once. The number of gene targets that are amplified and evaluated in each run limits the number of multi-well formats that are typically used for the analysis. The sequencing-by-synthesis method known as pyrosequencing uses luminometric detection of the pyrophosphate generated during nucleotide incorporation. The viability and dependability of this approach for the quantification of methylation at single CpGs inside PCR fragments have been demonstrated using bisulfite-converted DNA as a template. PCR is used to amplify a target region of up to 350 kb using primers corresponding to the DNA sequence that underwent bisulfite conversion. To facilitate the purification of single-stranded templates, one of the two amplification primers has a biotin residue at its 5'-terminus. An internal pyrosequencing primer complementary to the single-stranded template is hybridized to the template. The requirement to insert the primers into regions devoid of CpG sites places restrictions on this quantitative approach. Immunoprecipitation of highly methylated DNA sequences has been carried out using a 5-methylcytosine-specific antibody. The methylated fraction is captured from the reaction by utilizing Protein A/G beads after the fragments created by the sonication of the genomic DNA are treated with an anti-5mC antibody.

The majority of the constitutive heterochromatin is made up of areas with a lot of repetitive DNA elements, like transposable elements and satellite sequence clusters at telomeres, pericentric loci, and centromeres. Because they can insert into protein-coding genes, causing chromosome breakage and promoting genome rearrangement, transposable elements, which are prevalent in the human genome, are highly mutagenic. Because it prevents improper chromosome segregation, recombination, and DNA replication, the constitutive heterochromatin maintained by H3K9me3 is essential for maintaining genomic integrity. The rRNA gene and the transcriptionally active portion of euchromatin are the main sites of the facultative heterochromatin, which is instead mostly present at regulated loci where the chromatin state can vary in response to cellular signals.

Overall, a few aspects of genetic studies in life sciences from the old to the new have been discussed in the book.

REFERENCES

1. Abarca-Cabrera, L., Fraga-García, P., & Berensmeier, S. (2021). Bio-nano interactions: Binding proteins, polysaccharides, lipids and nucleic acids onto magnetic nanoparticles. *Biomaterials Research*, *25*(1), 12. https://doi.org/10.1186/s40824-021-00212-y
2. Alberts, B., Johnson, A., Lewis, J., Raff, M., Roberts, K., & Walter, P. (2002). From DNA to RNA. In *Molecular Biology of the Cell. 4th edition*. Garland Science. Retrieved from https://www.ncbi.nlm.nih.gov/books/NBK26887/
3. Anderson, J. C., & Schultz, P. G. (2003). Adaptation of an Orthogonal Archaeal Leucyl-tRNA and Synthetase Pair for Four-base, Amber, and Opal Suppression. *Biochemistry*, *42*(32), 9598–9608. https://doi.org/10.1021/bi034550w
4. Ansorge, W. J. (2009). Next-generation DNA sequencing techniques. *New Biotechnology*, *25*(4), 195–203. https://doi.org/10.1016/j.nbt.2008.12.009
5. Ashbolt, N. J. (2004). Microbial contamination of drinking water and disease outcomes in developing regions. *Toxicology*, *198*(1), 229–238. https://doi.org/10.1016/j.tox.2004.01.030
6. Ayala, F. J. (2007). Darwin's greatest discovery: Design without designer. *Proceedings of the National Academy of Sciences*, *104*(suppl_1), 8567–8573. https://doi.org/10.1073/pnas.0701072104

7. Ballaban-Gil, K., Rapin, I., Tuchman, R., & Shinnar, S. (1996). Longitudinal examination of the behavioral, language, and social changes in a population of adolescents and young adults with autistic disorder. *Pediatric Neurology*, *15*(3), 217–223. https://doi.org/10.1016/S0887-8994(96)00219-6
8. Barton, N., & Partridge, L. (2000). Limits to natural selection. *BioEssays*, *22*(12), 1075–1084. https://doi.org/10.1002/1521-1878(200012)22:12<1075::AID-BIES5>3.0.CO;2-M
9. Bearn, A. G. (1996). Inborn Errors of Metabolism: Garrod's Legacy. *Molecular Medicine*, *2*(3), 271–273. https://doi.org/10.1007/BF03401624
10. Bi, S., Zhang, H., Qiao, C., Sun, Y., & Liu, C. (2008). Studies of interaction of emodin and DNA in the presence of ethidium bromide by spectroscopic method. *Spectrochimica Acta Part A: Molecular and Biomolecular Spectroscopy*, *69*(1), 123–129. https://doi.org/10.1016/j.saa.2007.03.017
11. Birge, E. A. (2000). Transduction. In E. A. Birge (Ed.), *Bacterial and Bacteriophage Genetics* (pp. 293–313). New York, NY: Springer. https://doi.org/10.1007/978-1-4757-3258-0_9
12. Botstein, D., & Risch, N. (2003). Discovering genotypes underlying human phenotypes: Past successes for mendelian disease, future approaches for complex disease. *Nature Genetics*, *33*(3), 228–237. https://doi.org/10.1038/ng1090
13. Brown, A. S. (2012). Epidemiologic studies of exposure to prenatal infection and risk of schizophrenia and autism. *Developmental Neurobiology*, *72*(10), 1272–1276. https://doi.org/10.1002/dneu.22024
14. Brown, J. R., & Doolittle, W. F. (1997). Archaea and the prokaryote-to-eukaryote transition. *Microbiology and Molecular Biology Reviews*, *61*(4), 456–502. https://doi.org/10.1128/mmbr.61.4.456-502.1997
15. Buccelli, C., Della Casa, E., Paternoster, M., Niola, M., & Pieri, M. (2016). Gender differences in drug abuse in the forensic toxicological approach. *Forensic Science International*, *265*, 89–95. https://doi.org/10.1016/j.forsciint.2016.01.014
16. Busch, U. (2021). Progress in Physical X-Ray Research Until 1915. In U. Busch (Ed.), *Wilhelm Conrad Röntgen: A Shining Life for Science* (pp. 99–119). Cham: Springer International Publishing. https://doi.org/10.1007/978-3-030-72243-2_5

17. Campbell, I. D. (2002). The march of structural biology. *Nature Reviews Molecular Cell Biology*, *3*(5), 377–381. https://doi.org/10.1038/nrm800

18. Cappellini, M., & Fiorelli, G. (2008). Glucose-6-phosphate dehydrogenase deficiency. *The Lancet*, *371*(9606), 64–74. https://doi.org/10.1016/S0140-6736(08)60073-2

19. Carlson, M., & Botstein, D. (1982). Two differentially regulated mRNAs with different 5′ ends encode secreted and intracellular forms of yeast invertase. *Cell*, *28*(1), 145–154. https://doi.org/10.1016/0092-8674(82)90384-1

20. Charlton-Menys, V., & Durrington, P. N. (2008). Human cholesterol metabolism and therapeutic molecules. *Experimental Physiology*, *93*(1), 27–42. https://doi.org/10.1113/expphysiol.2006.035147

21. Chen, T., Tsujimoto, N., & Li, E. (2004). The PWWP Domain of Dnmt3a and Dnmt3b Is Required for Directing DNA Methylation to the Major Satellite Repeats at Pericentric Heterochromatin. Molecular and Cellular Biology, 24(20), 9048–9058. https://doi.org/10.1128/MCB.24.20.9048-9058.2004

22. Chuzhanova, N. A., Anassis, E. J., Ball, E. V., Krawczak, M., & Cooper, D. N. (2003). Meta-analysis of indels causing human genetic disease: Mechanisms of mutagenesis and the role of local DNA sequence complexity. *Human Mutation*, *21*(1), 28–44. https://doi.org/10.1002/humu.10146

23. Conner, B. N., Takano, T., Tanaka, S., Itakura, K., & Dickerson, R. E. (1982). The molecular structure of d(ICpCpGpG), a fragment of right-handed double helical A-DNA. *Nature*, *295*(5847), 294–299. https://doi.org/10.1038/295294a0

24. Cooper, R. (1984). A note on the biologic concept of race and its application in epidemiologic research. *American Heart Journal*, *108*(3, Part 2), 715–723. https://doi.org/10.1016/0002-8703(84)90662-8

25. Corcos, A. F., & Monaghan, F. V. (1990). Mendel's work and its rediscovery: A new perspective. *Critical Reviews in Plant Sciences*, *9*(3), 197–212. https://doi.org/10.1080/07352689009382287

26. Cornel, A. J., Q. Benedict, M., Salazar Rafferty, C., Howells, A. J., & Collins, Frank. H. (1997). Transient expression of the Drosophila melanogaster cinnabar gene rescues eye color in the white eye (WE) strain of Aedes aegypti. *Insect Biochemistry and Molecular Biology*,

27(12), 993–997. https://doi.org/10.1016/S0965-1748(97)00084-2

27. Dabrowska, K., Zembala, M., Boratynski, J., Switala-Jelen, K., Wietrzyk, J., Opolski, A., … Gorski, A. (2007). Hoc protein regulates the biological effects of T4 phage in mammals. *Archives of Microbiology*, *187*(6), 489–498. https://doi.org/10.1007/s00203-007-0216-y
28. Dahm, R. (2006). Alzheimer's discovery. *Current Biology*, *16*(21), R906–R910. https://doi.org/10.1016/j.cub.2006.09.056
29. Dahm, R. (2008). Discovering DNA: Friedrich Miescher and the early years of nucleic acid research. *Human Genetics*, *122*(6), 565–581. https://doi.org/10.1007/s00439-007-0433-0
30. Dahm, R., Gribbon, C., Quinlan, R. A., & Prescott, A. R. (1998). Changes in the nucleolar and coiled body compartments precede lamina and chromatin reorganization during fibre cell denucleation in the bovine lens. *European Journal of Cell Biology*, *75*(3), 237–246. https://doi.org/10.1016/S0171-9335(98)80118-0
31. Darden, L. (1977). William Bateson and the promise of Mendelism. *Journal of the History of Biology*, *10*(1), 87–106. https://doi.org/10.1007/BF00126096
32. Deichmann, U. (2016). Epigenetics: The origins and evolution of a fashionable topic. *Developmental Biology*, *416*(1), 249–254. https://doi.org/10.1016/j.ydbio.2016.06.005
33. Deng, W. P., & Nickoloff, J. A. (1992). Site-directed mutagenesis of virtually any plasmid by eliminating a unique site. *Analytical Biochemistry*, *200*(1), 81–88. https://doi.org/10.1016/0003-2697(92)90280-K
34. Doerfler, W. (2007). Human Adenovirus Type 12. Adenovirus Methods and Protocols, 131, 197–211. https://doi.org/10.1007/978-1-59745-277-9_14
35. Dokmanovic, M., & Marks, P. A. (2005). Prospects: Histone deacetylase inhibitors. *Journal of Cellular Biochemistry*, *96*(2), 293–304. https://doi.org/10.1002/jcb.20532
36. Durfee, T., Nelson, R., Baldwin, S., Plunkett, G., Burland, V., Mau, B., … Blattner, F. R. (2008). The Complete Genome Sequence of Escherichia coli DH10B: Insights into the Biology of a Laboratory Workhorse. *Journal of Bacteriology*, *190*(7), 2597–2606. https://doi.org/10.1128/jb.01695-07
37. Duric, N., Littrup, P., Babkin, A., Chambers, D., Azevedo, S., Kalinin,

A., ... Duncan, R. (2005). Development of ultrasound tomography for breast imaging: Technical assessment. *Medical Physics*, *32*(5), 1375–1386. https://doi.org/10.1118/1.1897463

38. Dyachenko, O. V., Tarlachkov, S. V., Marinitch, D. V., Shevchuk, T. V., & Buryanov, Y. I. (2014). Expression of exogenous DNA methyltransferases: Application in molecular and cell biology. *Biochemistry (Moscow)*, *79*(2), 77–87. https://doi.org/10.1134/S0006297914020011
39. Eisenberg, E., & Levanon, E. Y. (2003). Human housekeeping genes are compact. *Trends in Genetics*, *19*(7), 362–365. https://doi.org/10.1016/S0168-9525(03)00140-9
40. Farré-Maduell, E., & Casals-Pascual, C. (2019). The origins of gut microbiome research in Europe: From Escherich to Nissle. *Human Microbiome Journal*, *14*, 100065. https://doi.org/10.1016/j.humic.2019.100065
41. Fee, M. (2006). Racializing narratives: Obesity, diabetes and the "Aboriginal" thrifty genotype. *Social Science & Medicine*, *62*(12), 2988–2997. https://doi.org/10.1016/j.socscimed.2005.11.062
42. Flynt, L. (2020). 18F Sodium Fluoride: Tracer and Technique. In K. Kairemo & H. A. Macapinlac (Eds.), *Sodium Fluoride PET/CT in Clinical Use* (pp. 1–7). Cham: Springer International Publishing. https://doi.org/10.1007/978-3-030-23577-2_1
43. Fülöp, F., Martinek, T. A., & Tóth, G. K. (2006). Application of alicyclic β-amino acids in peptide chemistry. *Chemical Society Reviews*, *35*(4), 323–334. https://doi.org/10.1039/B501173F
44. Gao, J., Liu, H., & Kool, E. T. (2004). Expanded-Size Bases in Naturally Sized DNA: Evaluation of Steric Effects in Watson−Crick Pairing. *Journal of the American Chemical Society*, *126*(38), 11826–11831. https://doi.org/10.1021/ja048499a
45. Garrington, T. P., & Johnson, G. L. (1999). Organization and regulation of mitogen-activated protein kinase signaling pathways. *Current Opinion in Cell Biology*, *11*(2), 211–218. https://doi.org/10.1016/S0955-0674(99)80028-3
46. Godbole, K., Gayathri, P., Ghule, S., Sasirekha, B. V., Kanitkar-Damle, A., Memane, N., ... Yajnik, C. S. (2011). Maternal one-carbon metabolism, MTHFR and TCN2 genotypes and neural tube defects in India. *Birth Defects Research Part A: Clinical and Molecular*

Teratology, *91*(9), 848–856. https://doi.org/10.1002/bdra.20841

47. Grant-downton, R. T., & Dickinson, H. G. (2006). Epigenetics and its Implications for Plant Biology 2. The 'Epigenetic Epiphany': Epigenetics, Evolution and Beyond. Annals of Botany, 97(1), 11–27. https://doi.org/10.1093/aob/mcj001
48. Greenwood, C. M. T., Fujiwara, T. M., Boothroyd, L. J., Miller, M. A., Frappier, D., Fanning, E. A., … Morgan, K. (2000). Linkage of Tuberculosis to Chromosome 2q35 Loci, Including NRAMP1, in a Large Aboriginal Canadian Family. *The American Journal of Human Genetics*, *67*(2), 405–416. https://doi.org/10.1086/303012
49. Grummt, I., Kuhn, A., Bartsch, I., & Rosenbauer, H. (1986). A transcription terminator located upstream of the mouse rDNA initiation site affects rRNA synthesis. *Cell*, *47*(6), 901–911. https://doi.org/10.1016/0092-8674(86)90805-6
50. Haase, K., & Pelling, A. E. (2015). Investigating cell mechanics with atomic force microscopy. *Journal of The Royal Society Interface*, *12*(104), 20140970. https://doi.org/10.1098/rsif.2014.0970
51. Harvey, Z. H., Chakravarty, A. K., Futia, R. A., & Jarosz, D. F. (2020). A prion epigenetic switch establishes an active chromatin state. Cell, 180(5), 928-940.e14. https://doi.org/10.1016/j.cell.2020.02.014
52. Hazarika, P., & Russell, D. A. (2012). Advances in Fingerprint Analysis. *Angewandte Chemie International Edition*, *51*(15), 3524–3531. https://doi.org/10.1002/anie.201104313
53. Heide, K. M., & Solomon, E. P. (2006). Biology, childhood trauma, and murder: Rethinking justice. *International Journal of Law and Psychiatry*, *29*(3), 220–233. https://doi.org/10.1016/j.ijlp.2005.10.001
54. Hennah, W., Thomson, P., McQuillin, A., Bass, N., Loukola, A., Anjorin, A., … Porteous, D. (2009). DISC1 association, heterogeneity and interplay in schizophrenia and bipolar disorder. *Molecular Psychiatry*, *14*(9), 865–873. https://doi.org/10.1038/mp.2008.22
55. Herzstein, J., & Stave, G. M. (2016). Recombinant Organisms. In *Physical and Biological Hazards of the Workplace* (pp. 547–551). John Wiley & Sons, Ltd. https://doi.org/10.1002/9781119276531.ch31
56. Hojsgaard, D. H., Martínez, E. J., & Quarin, C. L. (2013). Competition between meiotic and apomictic pathways during ovule and seed development results in clonality. *New Phytologist*, *197*(1), 336–347. https://doi.org/10.1111/j.1469-8137.2012.04381.x

57. Houck, C. M., Rinehart, F. P., & Schmid, C. W. (1979). A ubiquitous family of repeated DNA sequences in the human genome. *Journal of Molecular Biology*, *132*(3), 289–306. https://doi.org/10.1016/0022-2836(79)90261-4

58. Hu, J., & Barrett, R. D. H. (2017). Epigenetics in natural animal populations. *Journal of Evolutionary Biology*, *30*(9), 1612–1632. https://doi.org/10.1111/jeb.13130

59. Humphreys, P. J. G., John Humphreys, John. (2014). *Electron Microscopy and Analysis* (3rd ed.). London: CRC Press. https://doi.org/10.1201/9781482289343

60. Jackson, M., Reinaldo Camacho, L., Gicquel, B., & Guilhot, C. (2001). Gene Replacement and Transposon Delivery Using the Negative Selection Marker sacB. In T. Parish & N. G. Stoker (Eds.), *Mycobacterium tuberculosis Protocols* (pp. 59–75). Totowa, NJ: Humana Press. https://doi.org/10.1385/1-59259-147-7:059

61. Jaskolski, M., Dauter, Z., & Wlodawer, A. (2014). A brief history of macromolecular crystallography, illustrated by a family tree and its Nobel fruits. *The FEBS Journal*, *281*(18), 3985–4009. https://doi.org/10.1111/febs.12796

62. Jenkin, J. (2001). A Unique Partnership: William and Lawrence Bragg and the 1915 Nobel Prize in Physics. *Minerva*, *39*(4), 373–392. https://doi.org/10.1023/A:1012783802528

63. Jensen, R. E., Parks, M. M., Mann, B. W., Maison, K., & Krall, M. A. (2019). Mapping Nature's scientist: The posthumous demarcation of Rosalind Franklin's crystallographic data. *Quarterly Journal of Speech*, *105*(3), 297–318. https://doi.org/10.1080/00335630.2019.1629000

64. Jia, D., Jurkowska, R. Z., Zhang, X., Jeltsch, A., & Cheng, X. (2007). Structure of Dnmt3a bound to Dnmt3L suggests a model for de novo DNA methylation. Nature, 449(7159), 248–251. https://doi.org/10.1038/nature06146

65. Kaikkonen, M. U., Lam, M. T. Y., & Glass, C. K. (2011). Non-coding RNAs as regulators of gene expression and epigenetics. Cardiovascular Research, 90(3), 430–440. https://doi.org/10.1093/cvr/cvr097

66. Kalinin, S. V., Rodriguez, B. J., Jesse, S., Karapetian, E., Mirman, B., Eliseev, E. A., & Morozovska, A. N. (2007). Nanoscale Electromechanics of Ferroelectric and Biological Systems: A New

Dimension in Scanning Probe Microscopy. *Annual Review of Materials Research*, *37*(1), 189–238. https://doi.org/10.1146/annurev.matsci.37.052506.084323

67. Kantardjieff, K. A., & Rupp, B. (2003). Matthews coefficient probabilities: Improved estimates for unit cell contents of proteins, DNA, and protein–nucleic acid complex crystals. *Protein Science*, *12*(9), 1865–1871. https://doi.org/10.1110/ps.0350503
68. Katzmann, E., Müller, F. D., Lang, C., Messerer, M., Winklhofer, M., Plitzko, J. M., & Schüler, D. (2011). Magnetosome chains are recruited to cellular division sites and split by asymmetric septation. *Molecular Microbiology*, *82*(6), 1316–1329. https://doi.org/10.1111/j.1365-2958.2011.07874.x
69. Kearney, H. M., Thorland, E. C., Brown, K. K., Quintero-Rivera, F., & South, S. T. (2011). American College of Medical Genetics standards and guidelines for interpretation and reporting of postnatal constitutional copy number variants. *Genetics in Medicine*, *13*(7), 680–685. https://doi.org/10.1097/GIM.0b013e3182217a3a
70. Kirk, D. L. (2005). A twelve-step program for evolving multicellularity and a division of labor. *BioEssays*, *27*(3), 299–310. https://doi.org/10.1002/bies.20197
71. Klump, H. (1987). Codification and evolution of experimentally observed specific recognition sites for restriction enzymes on DNA. *Biosystems*, *21*(1), 33–49. https://doi.org/10.1016/0303-2647(87)90005-0
72. Kudlicki, W., Coffman, A., Kramer, G., & Hardesty, B. (1997). Ribosomes and ribosomal RNA as chaperones for folding of proteins. *Folding and Design*, *2*(2), 101–108. https://doi.org/10.1016/S1359-0278(97)00014-X
73. Kuo, M.-H., & Allis, C. D. (1999). In Vivo Cross-Linking and Immunoprecipitation for Studying Dynamic Protein:DNA Associations in a Chromatin Environment. *Methods*, *19*(3), 425–433. https://doi.org/10.1006/meth.1999.0879
74. Langanke, K., Martínez-Pinedo, G., Sampaio, J. M., Dean, D. J., Hix, W. R., Messer, O. E. B., … Rampp, M. (2003). Electron Capture Rates on Nuclei and Implications for Stellar Core Collapse. *Physical Review Letters*, *90*(24), 241102. https://doi.org/10.1103/PhysRevLett.90.241102

75. Lederberg, J. (1994). The Transformation of Genetics by DNA: An Anniversary Celebration of Avery, Macleod and Mccarty (1944). *Genetics*, *136*(2), 423–426. Retrieved from https://www.ncbi.nlm.nih.gov/pmc/articles/PMC1205797/
76. Leitão, E., Beygo, J., Zeschnigk, M., Klein-Hitpass, L., Bargull, M., Rahmann, S., & Horsthemke, B. (2018). Locus-Specific DNA Methylation Analysis by Targeted Deep Bisulfite Sequencing. In A. Jeltsch & M. G. Rots (Eds.), *Epigenome Editing: Methods and Protocols* (pp. 351–366). New York, NY: Springer. https://doi.org/10.1007/978-1-4939-7774-1_19
77. Lerminiaux, N. A., & Cameron, A. D. S. (2019). Horizontal transfer of antibiotic resistance genes in clinical environments. *Canadian Journal of Microbiology*, *65*(1), 34–44. https://doi.org/10.1139/cjm-2018-0275
78. Li, A., Qu, G., Sun, Z., & Reetz, M. T. (2019). Statistical Analysis of the Benefits of Focused Saturation Mutagenesis in Directed Evolution Based on Reduced Amino Acid Alphabets. *ACS Catalysis*, *9*(9), 7769–7778. https://doi.org/10.1021/acscatal.9b02548
79. Liu, G., Grant, W. M., Persky, D., Latham, V. M., Singer, R. H., & Condeelis, J. (2002). Interactions of Elongation Factor 1α with F-Actin and β-Actin mRNA: Implications for Anchoring mRNA in Cell Protrusions. *Molecular Biology of the Cell*, *13*(2), 579–592. https://doi.org/10.1091/mbc.01-03-0140
80. Loscalzo, J., & Handy, D. E. (2014). Epigenetic modifications: Basic mechanisms and role in cardiovascular disease (2013 Grover Conference series). Pulmonary Circulation, 4(2), 169–174. https://doi.org/10.1086/675979
81. Loughlin, R., Wilbur, J. D., McNally, F. J., Nédélec, F. J., & Heald, R. (2011). Katanin Contributes to Interspecies Spindle Length Scaling in Xenopus. *Cell*, *147*(6), 1397–1407. https://doi.org/10.1016/j.cell.2011.11.014
82. Magno, L., Lessard, C. B., Martins, M., Lang, V., Cruz, P., Asi, Y., Whiting, P. J. (2019). Alzheimer's disease phospholipase C-gamma-2 (PLCG2) protective variant is a functional hypermorph. *Alzheimer's Research & Therapy*, *11*(1), 16. https://doi.org/10.1186/s13195-019-0469-0
83. Mahmood, N., Cheishvili, D., Arakelian, A., Tanvir, I., Khan, H. A., Pépin, A.-S., Rabbani, S. A. (2017). Methyl donor S-adenosylmethionine (SAM) supplementation attenuates breast

cancer growth, invasion, and metastasis in vivo; therapeutic and chemopreventive applications. Oncotarget, 9(4), 5169–5183. https://doi.org/10.18632/oncotarget.23704

84. Mardis, E. R. (2011). A decade's perspective on DNA sequencing technology. *Nature*, *470*(7333), 198–203. https://doi.org/10.1038/nature09796

85. Marmur, J. (1961). A procedure for the isolation of deoxyribonucleic acid from micro-organisms. *Journal of Molecular Biology*, *3*(2), 208-IN1. https://doi.org/10.1016/S0022-2836(61)80047-8

86. Maron, D. J., Fair, J. M., & Haskell, W. L. (1991). Saturated fat intake and insulin resistance in men with coronary artery disease. The Stanford Coronary Risk Intervention Project Investigators and Staff. *Circulation*, *84*(5), 2020–2027. https://doi.org/10.1161/01.CIR.84.5.2020

87. Maxam, A. M., & Gilbert, W. (1980). [57] Sequencing end-labeled DNA with base-specific chemical cleavages. In *Nucleic Acids Part I*: *Vol. 65*. *Methods in Enzymology* (pp. 499–560). Academic Press. https://doi.org/10.1016/S0076-6879(80)65059-9

88. McGrew, J. T., Goetsch, L., Byers, B., & Baum, P. (1992). Requirement for ESP1 in the nuclear division of Saccharomyces cerevisiae. *Molecular Biology of the Cell*, *3*(12), 1443–1454. https://doi.org/10.1091/mbc.3.12.1443

89. Merikangas, K. R., & McClair, V. L. (2012). Epidemiology of substance use disorders. *Human Genetics*, *131*(6), 779–789. https://doi.org/10.1007/s00439-012-1168-0

90. Méthot, P.-O. (2016). Bacterial Transformation and the Origins of Epidemics in the Interwar Period: The Epidemiological Significance of Fred Griffith's "Transforming Experiment." *Journal of the History of Biology*, *49*(2), 311–358. https://doi.org/10.1007/s10739-015-9415-6

91. Meyvis, T. K. L., De Smedt, S. C., Van Oostveldt, P., & Demeester, J. (1999). Fluorescence Recovery After Photobleaching: A Versatile Tool for Mobility and Interaction Measurements in Pharmaceutical Research. *Pharmaceutical Research*, *16*(8), 1153–1162. https://doi.org/10.1023/A:1011924909138

92. Mikami, T., Mitomi, H., Hara, A., Yanagisawa, N., Yoshida, T., Tsuruta, O., & Okayasu, I. (2000). Decreased expression of CD44, alpha-catenin, and deleted colon carcinoma and altered expression of beta-catenin in ulcerative colitis-associated dysplasia and carcinoma,

as compared with sporadic colon neoplasms. *Cancer*, *89*(4), 733–740. https://doi.org/10.1002/1097-0142(20000815)89:4<733::AID-CNCR3>3.0.CO;2-#

93. Mori, S., & Barker, P. B. (1999). Diffusion magnetic resonance imaging: Its principle and applications. *The Anatomical Record*, *257*(3), 102–109. https://doi.org/10.1002/(SICI)1097-0185(19990615)257:3<102::AID-AR7>3.0.CO;2-6
94. Motorin, Y., & Helm, M. (2011). RNA nucleotide methylation. *WIREs RNA*, *2*(5), 611–631. https://doi.org/10.1002/wrna.79
95. Moysich, K. B., Menezes, R. J., & Michalek, A. M. (2002). Chernobyl-related ionising radiation exposure and cancer risk: An epidemiological review. *The Lancet Oncology*, *3*(5), 269–279. https://doi.org/10.1016/S1470-2045(02)00727-1
96. Mozo, T., Dewar, K., Dunn, P., Ecker, J. R., Fischer, S., Kloska, S., … Altmann, T. (1999). A complete BAC-based physical map of the Arabidopsis thaliana genome. *Nature Genetics*, *22*(3), 271–275. https://doi.org/10.1038/10334
97. Muller, H. J., & Altenburg, E. (1930). The Frequency of Translocations Produced by X-Rays in Drosophila. *Genetics*, *15*(4), 283–311. Retrieved from https://www.ncbi.nlm.nih.gov/pmc/articles/PMC1201066/
98. Musiol-Kroll, E. M., Tocchetti, A., Sosio, M., & Stegmann, E. (2019). Challenges and advances in genetic manipulation of filamentous actinomycetes – the remarkable producers of specialized metabolites. *Natural Product Reports*, *36*(9), 1351–1369. https://doi.org/10.1039/C9NP00029A
99. Myers, S. P., Timken, M. D., Piucci, M. L., Sims, G. A., Greenwald, M. A., Weigand, J. J., … Buoncristiani, M. R. (2011). Searching for first-degree familial relationships in California's offender DNA database: Validation of a likelihood ratio-based approach. *Forensic Science International: Genetics*, *5*(5), 493–500. https://doi.org/10.1016/j.fsigen.2010.10.010
100. Ndagi, U., A. Falaki, A., Abdullahi, M., M. Lawal, M., & E. Soliman, M. (2020). Antibiotic resistance: Bioinformatics-based understanding as a functional strategy for drug design. *RSC Advances*, *10*(31), 18451–18468. https://doi.org/10.1039/D0RA01484B
101. Newman, S. W. (1999). The Medial Extended Amygdala in Male Reproductive Behavior A Node in the Mammalian Social Behavior

Network. *Annals of the New York Academy of Sciences*, *877*(1), 242–257. https://doi.org/10.1111/j.1749-6632.1999.tb09271.x

102. Nyholt, D. R., Morley, K. I., Ferreira, M. A. R., Medland, S. E., Boomsma, D. I., Heath, A. C., … Martin, N. G. (2005). Genomewide Significant Linkage to Migrainous Headache on Chromosome 5q21. *The American Journal of Human Genetics*, *77*(3), 500–512. https://doi.org/10.1086/444510
103. Oakes, C. C., La Salle, S., Robaire, B., & Trasler, J. M. (2006). Evaluation of a Quantitative DNA Methylation Analysis Technique using Methylation-Sensitive/Dependent Restriction Enzymes and Real-Time PCR. *Epigenetics*, *1*(3), 146–152. https://doi.org/10.4161/epi.1.3.3392
104. Okamura, M., Inose, H., & Masuda, S. (2015). RNA Export through the NPC in Eukaryotes. *Genes*, *6*(1), 124–149. https://doi.org/10.3390/genes6010124
105. Patrono, C. (2023). Fifty years with aspirin and platelets. *British Journal of Pharmacology*, *180*(1), 25–43. https://doi.org/10.1111/bph.15966
106. Patwardhan, D. (2022). Mendelian Principle of Inheritance. In D. Kar & S. Sarkar (Eds.), *Genetics Fundamentals Notes* (pp. 53–83). Singapore: Springer Nature. https://doi.org/10.1007/978-981-16-7041-1_2
107. Peterson, R. F., & Treagust, D. F. (1989, June 1). Grade-12 students' misconceptions of covalent bonding and structure [Research-article]. https://doi.org/10.1021/ed066p459
108. Phillips, P. C. (2008). Epistasis—The essential role of gene interactions in the structure and evolution of genetic systems. *Nature Reviews Genetics*, *9*(11), 855–867. https://doi.org/10.1038/nrg2452
109. Pickrell, J. K., Berisa, T., Liu, J. Z., Ségurel, L., Tung, J. Y., & Hinds, D. A. (2016). Detection and interpretation of shared genetic influences on 42 human traits. *Nature Genetics*, *48*(7), 709–717. https://doi.org/10.1038/ng.3570
110. PORTIN, P. (2014). The birth and development of the DNA theory of inheritance: Sixty years since the discovery of the structure of DNA. *Journal of Genetics*, *93*(1), 293–302. https://doi.org/10.1007/s12041-014-0337-4
111. Potts, P. J., Ellis, A. T., Kregsamer, P., Streli, C., Vanhoof, C., West, M., & Wobrauschek, P. (2005). Atomic spectrometry update. X-Ray

fluorescence spectrometry. *Journal of Analytical Atomic Spectrometry, 20*(10), 1124–1154. https://doi.org/10.1039/B511542F

112. Prasher, D. C., Eckenrode, V. K., Ward, W. W., Prendergast, F. G., & Cormier, M. J. (1992). Primary structure of the Aequorea victoria green-fluorescent protein. *Gene, 111*(2), 229–233. https://doi.org/10.1016/0378-1119(92)90691-H
113. Qureshi, I. A., & Mehler, M. F. (2018). Epigenetic mechanisms underlying nervous system diseases. Handbook of Clinical Neurology, 147, 43–58. https://doi.org/10.1016/B978-0-444-63233-3.00005-1
114. Raju, T. N. (1999). The Nobel chronicles. 1958: George Wells Beadle (1903-89), Edward Lawrie Tatum (1909-75) and Joshua Lederberg (b 1925). *Lancet (London, England), 353*(9169), 2082. https://doi.org/10.1016/s0140-6736(05)77919-8
115. Randolph, M. K., & Zhao, W. (2015). Genome editing and stem cell therapy pave the path for new treatment of sickle-cell disease. *Stem Cell Investigation, 2*, 22. https://doi.org/10.3978/j.issn.2306-9759.2015.11.02
116. Rende, R. D., Plomin, R., & Vandenberg, S. G. (1990). Who discovered the twin method? *Behavior Genetics, 20*(2), 277–285. https://doi.org/10.1007/BF01067795
117. Richmond, M. L. (2006). "The 'Domestication' of Heredity: The Familial Organization of Geneticists at Cambridge University, 1895–1910." *Journal of the History of Biology, 39*(3), 565–605. https://doi.org/10.1007/s10739-004-5431-7
118. Rogan, P. K., Faux, B. M., & Schneider, T. D. (1998). Information analysis of human splice site mutations. *Human Mutation, 12*(3), 153–171. https://doi.org/10.1002/(SICI)1098-1004(1998)12:3<153::AID-HUMU3>3.0.CO;2-I
119. Roos, R. A. (2010). Huntington's disease: A clinical review. *Orphanet Journal of Rare Diseases, 5*(1), 40. https://doi.org/10.1186/1750-1172-5-40
120. Russo, G., De Franceschi, L., Colombatti, R., Rigano, P., Perrotta, S., Voi, V., … Forni, G. L. (2019). Current challenges in the management of patients with sickle cell disease – A report of the Italian experience. *Orphanet Journal of Rare Diseases, 14*(1), 120. https://doi.org/10.1186/s13023-019-1099-0
121. Schubert, R., Trenholm, S., Balint, K., Kosche, G., Cowan, C. S.,

Mohr, M. A., … Roska, B. (2018). Virus stamping for targeted single-cell infection in vitro and in vivo. *Nature Biotechnology*, *36*(1), 81–88. https://doi.org/10.1038/nbt.4034

122. Schüpbach, T. (2019). Genetic Screens to Analyze Pattern Formation of Egg and Embryo in Drosophila: A Personal History. *Annual Review of Genetics*, *53*(1), 1–18. https://doi.org/10.1146/annurev-genet-112618-043708
123. Schwarzbach, E., Smýkal, P., Dostál, O., Jarkovská, M., & Valová, S. (2014). Gregor J. Mendel—Genetics founding father. *Czech Journal of Genetics and Plant Breeding*, *50*(2), 43–51. https://doi.org/10.17221/54/2014-CJGPB
124. Selvin, P. R. (1995). [13] Fluorescence resonance energy transfer. In *Biochemical Spectroscopy*: *Vol. 246. Methods in Enzymology* (pp. 300–334). Academic Press. https://doi.org/10.1016/0076-6879(95)46015-2
125. Shimomura, O. (2005). Discovery of Green Fluorescent Protein. In *Green Fluorescent Protein* (pp. 1–13). John Wiley & Sons, Ltd. https://doi.org/10.1002/0471739499.ch1
126. Sivanandam, S. N., & Deepa, S. N. (2008). Genetic Algorithms. In S. N. Sivanandam & S. N. Deepa (Eds.), *Introduction to Genetic Algorithms* (pp. 15–37). Berlin, Heidelberg: Springer. https://doi.org/10.1007/978-3-540-73190-0_2
127. Skulason, S., & Smith, T. B. (1995). Resource polymorphisms in vertebrates. *Trends in Ecology & Evolution*, *10*(9), 366–370. https://doi.org/10.1016/S0169-5347(00)89135-1
128. Smith, Z. D., Gu, H., Bock, C., Gnirke, A., & Meissner, A. (2009). High-throughput bisulfite sequencing in mammalian genomes. *Methods*, *48*(3), 226–232. https://doi.org/10.1016/j.ymeth.2009.05.003
129. Stahl, F. W., & Murray, N. E. (1966). The Evolution of Gene Clusters and Genetic Circularity in Microorganisms. *Genetics*, *53*(3), 569–576. Retrieved from https://www.ncbi.nlm.nih.gov/pmc/articles/PMC1211040/
130. Stern, C. (1936). Somatic Crossing over and Segregation in Drosophila Melanogaster. *Genetics*, *21*(6), 625–730. Retrieved from https://www.ncbi.nlm.nih.gov/pmc/articles/PMC1208727/
131. Sternberg, N. L., & Maurer, R. (1991). [2] Bacteriophage-mediated generalized transduction in Escherichia coli and Salmonella typhimurium. In *Bacterial Genetic Systems*: *Vol. 204. Methods in*

Enzymology (pp. 18–43). Academic Press. https://doi.org/10.1016/0076-6879(91)04004-8

132. Sussman, M. (2002). Duchenne Muscular Dystrophy. *JAAOS - Journal of the American Academy of Orthopaedic Surgeons*, *10*(2), 138. Retrieved from https://journals.lww.com/jaaos/Abstract/2002/03000/Duchenne_Muscular_Dystrophy.9.aspx
133. Taby, R., & Issa, J.-P. J. (2010). Cancer Epigenetics. *CA: A Cancer Journal for Clinicians*, *60*(6), 376–392. https://doi.org/10.3322/caac.20085
134. Tolonen, N., Doglio, L., Schleich, S., & Locker, J. K. (2001). Vaccinia Virus DNA Replication Occurs in Endoplasmic Reticulum-enclosed Cytoplasmic Mini-Nuclei. *Molecular Biology of the Cell*, *12*(7), 2031–2046. https://doi.org/10.1091/mbc.12.7.2031
135. Tomkinson, A. E., Vijayakumar, S., Pascal, J. M., & Ellenberger, T. (2006). DNA Ligases: Structure, Reaction Mechanism, and Function. *Chemical Reviews*, *106*(2), 687–699. https://doi.org/10.1021/cr040498d
136. Tan, L., & Shi, Y. G. (2012). Tet family proteins and 5-hydroxymethylcytosine in development and disease. Development (Cambridge, England), 139(11), 1895–1902. https://doi.org/10.1242/dev.070771
137. Travers, A. (1993). The mechanism of RNA chain initiation. In A. Travers (Ed.), *DNA-Protein Interactions* (pp. 87–108). Dordrecht: Springer Netherlands. https://doi.org/10.1007/978-94-011-1480-6_4
138. Trouche, S., Pompili, M. N., & Girardeau, G. (2020). The role of sleep in emotional processing: Insights and unknowns from rodent research. *Current Opinion in Physiology*, *15*, 230–237. https://doi.org/10.1016/j.cophys.2020.04.003
139. Tsui, L.-C. (1995). The Cystic Fibrosis Transmembrane Conductance Regulator Gene. *American Journal of Respiratory and Critical Care Medicine*, *151*(3_pt_2), S47–S53. https://doi.org/10.1164/ajrccm/151.3_Pt_2.S47
140. Velculescu, V. E., Zhang, L., Zhou, W., Vogelstein, J., Basrai, M. A., Bassett, D. E., … Kinzler, K. W. (1997). Characterization of the Yeast Transcriptome. *Cell*, *88*(2), 243–251. https://doi.org/10.1016/S0092-8674(00)81845-0
141. Vereecke, D., Zhang, Y., Francis, I. M., Lambert, P. Q., Venneman, J.,

Stamler, R. A., … Randall, J. J. (2020). Functional Genomics Insights Into the Pathogenicity, Habitat Fitness, and Mechanisms Modifying Plant Development of Rhodococcus sp. PBTS1 and PBTS2. *Frontiers in Microbiology*, *11*. Retrieved from https://www.frontiersin.org/articles/10.3389/fmicb.2020.00014

142. Villanueva, J., Philip, J., Entenberg, D., Chaparro, C. A., Tanwar, M. K., Holland, E. C., & Tempst, P. (2004). Serum Peptide Profiling by Magnetic Particle-Assisted, Automated Sample Processing and MALDI-TOF Mass Spectrometry. *Analytical Chemistry*, *76*(6), 1560–1570. https://doi.org/10.1021/ac0352171
143. Wang, D., Zhao, C., Xu, F., Zhang, A., Jin, M., Zhang, K., … Huang, G. (2021). Cisplatin-resistant NSCLC cells induced by hypoxia transmit resistance to sensitive cells through exosomal PKM2. *Theranostics*, *11*(6), 2860–2875. https://doi.org/10.7150/thno.51797
144. Wang, X., Zhao, L., Wu, X., Luo, H., Wu, D., Zhang, M., … Zeng, Z. (2021). Development of a simplified and inexpensive RNA depletion method for plasmid DNA purification using size selection magnetic beads (SSMBs). *Genes & Diseases*, *8*(3), 298–306. https://doi.org/10.1016/j.gendis.2020.04.013
145. Warburton, P. E., & Willard, H. F. (1990). Genomic analysis of sequence variation in tandemly repeated DNA: Evidence for localized homogeneous sequence domains within arrays of α-satellite DNA. *Journal of Molecular Biology*, *216*(1), 3–16. https://doi.org/10.1016/S0022-2836(05)80056-7
146. Weiss, P. S. (2008). 2008 Nobel Prize in Chemistry: Green Fluorescent Protein, Its Variants and Implications. *ACS Nano*, *2*(10), 1977–1977. https://doi.org/10.1021/nn800671h
147. Wilson, M. R., DiZinno, J. A., Polanskey, D., Replogle, J., & Budowle, B. (1995). Validation of mitochondrial DNA sequencing for forensic casework analysis. *International Journal of Legal Medicine*, *108*(2), 68–74. https://doi.org/10.1007/BF01369907
148. Wogan, G. N., Hecht, S. S., Felton, J. S., Conney, A. H., & Loeb, L. A. (2004). Environmental and chemical carcinogenesis. *Seminars in Cancer Biology*, *14*(6), 473–486. https://doi.org/10.1016/j.semcancer.2004.06.010
149. Wüthrich, K. (2001). The way to NMR structures of proteins. *Nature Structural Biology*, *8*(11), 923–925. https://doi.org/10.1038/nsb1101-923

150. Yang, J., Brown, M. S., Liang, G., Grishin, N. V., & Goldstein, J. L. (2008). Identification of the Acyltransferase that Octanoylates Ghrelin, an Appetite-Stimulating Peptide Hormone. *Cell*, *132*(3), 387–396. https://doi.org/10.1016/j.cell.2008.01.017

151. Yother, J., McDaniel, L. S., & Briles, D. E. (1986). Transformation of encapsulated Streptococcus pneumoniae. *Journal of Bacteriology*, *168*(3), 1463–1465. https://doi.org/10.1128/jb.168.3.1463-1465.1986

152. Zhou, Q., Wang, M., Yuan, Y., Wang, X., Fu, R., Wan, H., … Zhou, Q. (2016). Complete Meiosis from Embryonic Stem Cell-Derived Germ Cells In Vitro. *Cell Stem Cell*, *18*(3), 330–340. https://doi.org/10.1016/j.stem.2016.01.017

153. Zubakov, D., Chamier-Ciemińska, J., Kokmeijer, I., Maciejewska, A., Martínez, P., Pawłowski, R., … Kayser, M. (2018). Introducing novel type of human DNA markers for forensic tissue identification: DNA copy number variation allows the detection of blood and semen. *Forensic Science International: Genetics*, *36*, 112–118. https://doi.org/10.1016/j.fsigen.2018.06.021

INDEX

N

O

P

Q

R

S